Studies in Fuzziness and Soft Computing

Volume 426

The series "Studies in Fuzziness and Soft Computing" contains publications on various topics in the area of soft computing, which include fuzzy sets, rough sets, neural networks, evolutionary computation, probabilistic and evidential reasoning, multi-valued logic, and related fields. The publications within "Studies in Fuzziness and Soft Computing" are primarily monographs and edited volumes. They cover significant recent developments in the field, both of a foundational and applicable character. An important feature of the series is its short publication time and world-wide distribution. This permits a rapid and broad dissemination of research results.

Indexed by SCOPUS, DBLP, WTI Frankfurt eG, zbMATH, SCImago.

All books published in the series are submitted for consideration in Web of Science.

Tanveen Kaur Bhatia · Amit Kumar ·
Srimantoorao S. Appadoo

More-for-Less Solutions in Fuzzy Transportation Problems

Springer

Tanveen Kaur Bhatia
School of Mathematics
Thapar Institute of Engineering and Technology
Patiala, Punjab, India

Amit Kumar
School of Mathematics
Thapar Institute of Engineering and Technology
Patiala, Punjab, India

Srimantoorao S. Appadoo
Department of Supply Chain Management
Asper School of Business
University of Manitoba
Winnipeg, MB, Canada

ISSN 1434-9922 ISSN 1860-0808 (electronic)
Studies in Fuzziness and Soft Computing
ISBN 978-3-031-30336-4 ISBN 978-3-031-30337-1 (eBook)
https://doi.org/10.1007/978-3-031-30337-1

This Springer imprint is published by the registered company Springer Nature Switzerland AG
The registered company address is: Gewerbestrasse 11, 6330 Cham, Switzerland

Dedicated
to
Parents
and
God

Acknowledgements

The authors would like to thank the Series Editor **Prof. Janusz Kacprzyk** for his valuable suggestions.

Dr. Tanveen Kaur Bhatia would like to express her heartfelt gratitude to her mother Mrs. Ravinderjeet Kaur and her father Mr. Tarvinderjeet Singh Bhatia for their unconditional love, support and blessings. She gratefully acknowledges the patience and love of her sibling Amritjeet Singh Bhatia that helped her to overcome the difficulties encountered during her life.

Dr. Amit Kumar would like to acknowledge the adolescent inner blessings of Mehar (daughter of his cousin Dr. Parmpreet Kaur). He believes that Mata Vaishno Devi has appeared on the earth in the form of Mehar, and without Mehar's blessings, it would not have been possible to think the ideas presented in this thesis.

Last but not least, both the authors are ever grateful to Almighty God. Thank you God for the numerous blessings bestowed upon us in every aspect of our life.

Contents

About the Authors

Dr. Tanveen Kaur Bhatia completed her Ph.D. thesis in School of Mathematics at Thapar Institute of Engineering and Technology, Patiala. She has received her M.Sc. degree in Mathematics and Computing from Thapar Institute of Engineering and Technology, Patiala. She has also received an institute medal for securing highest cumulative grade point average in Master of Science in Mathematics and Computing examination. She has received her B.Sc. degree in Non-Medical from Multani Mal Modi College, Patiala. Her main research interest is in fuzzy optimization. She has published eight research papers in SCIE indexed journals, Scopus indexed journal and Scopus indexed book series and delivered a contributed talk titled "A fuzzy logic based approach to solve interval multi-objective non-linear transportation problem: Suggested modifications" in Virtual International Conference entitled "Soft Computing, Optimization Theory and Applications" organized at BIT Mesra, Ranchi, India, during March 26–27, 2021.

Dr. Amit Kumar is working as Associate Professor in School of Mathematics, Thapar Institute of Engineering and Technology, Patiala. He has completed his M.Sc. and Ph.D. from Indian Institute of Technology Roorkee. Till date, he has guided 21 Ph.D. students and 17 M.Sc. students. He is co-author of approximate 74 research papers published in SCI/SCIE listed Journals. He is also co-author of four monographs published in the Springer book series "Studies in Fuzziness and Soft Computing".

Prof. Srimantoorao S. Appadoo is Professor of Management Science in the Department of Supply Chain Management at the University of Manitoba, Canada. His research work is published in esteemed international journals and referred proceedings. Some of his articles appeared on the Top 25 Hottest Articles on Science Direct Elsevier and were among the most cited articles on Elsevier. The National Sciences and Engineering Research Council of Canada (NSERC) Discovery Grant funds his research. Professor Appadoo is also the recipient of various research awards; students appreciate his teaching methodology and consistently rank his classes and instruction well above average. He serves on departmental, faculty and university committees and contributes extensively to the professional community through review processes, sitting as conference chair, academic reviewer and other scholarly activities. He has served as an external examiner for various M.Sc. and Ph.D. thesis committees.

Chapter 1
Introduction

In this chapter, firstly, the origin of more-for-less solutions of transportation problems is discussed. Then, a brief review of existing methods to find more-for-less solutions of transportation problems is discussed. Finally, the work done in each chapter is discussed in a brief manner.

1.1 Origin of More-For-Less Solutions of Transportation Problems

Transportation problem [13] is a well-known topic of Operations Research. Several methods are proposed in the literature to find an optimal solution of transportation problems. Szwarc [21] proposed the concept of more-for-less solution of balanced transportation problem which is defined as

> Under some specific conditions, by increasing the availability of the product at a specific source and the demand of the product at a specific destination, such new values of the existing optimal basic variables can be obtained corresponding to which (i) the total transported units of the product will be more than the optimal transported units of the product (ii) the total transportation cost will be less than the total optimal transportation cost. The set of new values of the existing optimal basic variables and existing optimal non-basic variables is known as a more-for-less solution of a transportation problem.

Szwarc [21] also proposed a condition under which more-for-less solution will exist for a balanced transportation problem.

T. K. Bhatia et al., *More-for-Less Solutions in Fuzzy Transportation Problems*, Studies in Fuzziness and Soft Computing 426,
https://doi.org/10.1007/978-3-031-30337-1_1

1.2 Literature Review

Inspired by the work of Szwarc [21], several methods are proposed in the literature to find more-for-less solutions of single-objective transportation problems [1–8, 11, 12, 14–16, 18–23].

In this section, a brief review of some existing methods to find more-for-less solutions of transportation problems is discussed.

Verma and Puri [22] proposed a condition under which at least one more-for-less solution will exist for a balanced crisp linear fractional transportation problem. Verma and Puri [22] also proposed a method to find a more-for-less solution of balanced crisp linear fractional transportation problems.

Arsham [7] proposed a perturbation analysis-based post optimality analysis method to find a more-for-less solution of balanced crisp transportation problems.

Gupta et al. [11] proposed a condition under which at least one more-for-less solution will exist for a crisp linear fractional transportation problem with mixed constraints. Gupta et al. [11] also proposed a method to find a more-for-less solution of crisp linear fractional transportation problems with mixed constraints.

Gupta and Puri [12] proposed a condition under which at least one more-for-less solution will exist for a crisp minimal cost flow problem with bounded variables. Gupta and Puri [12] also proposed a method to find a more-for-less solution of crisp minimal cost flow problems with bounded variables.

Adlakha and Kowalski [1] proposed an absolute point-based method to find a more-for-less solution of balanced crisp transportation problems.

Adlakha and Kowalski [2] generalized the existing method [1] to find a more-for-less solution of balanced fixed charge crisp transportation problems.

Adlakha and Kowalski [3] pointed out that under a specific condition, the existing method [1] fails to find a correct more-for-less solution of balanced crisp transportation problems. To overcome this limitation, Adlakha and Kowalski [3] modified the existing method [1] to find a correct more-for-less solution of balanced crisp transportation problems.

Adlakha and Kowalski [4] pointed out a limitation of the existing methods [2, 3]. Also, to overcome the limitation, Adlakha and Kowalski [4] proposed a heuristic algorithm to find a more-for-less solution of balanced crisp transportation problems as well as balanced fixed charge crisp transportation problems.

Deineko et al. [10] pointed out that as all existing conditions for the non-existence of a more-for-less solution are dependent upon the dual variables of the linear programming problem corresponding to the considered crisp transportation problem. Therefore, much computational efforts are required to verify the existing conditions for the non-existence of a more-for-less solution. To reduce the computational efforts, Deineko et al. [10] proposed such a condition for the non-existence of more-for-less solution which is only dependent upon the cost matrix.

Adlakha et al. [5] proposed a heuristic method to find a more-for-less solution of crisp transportation problems with mixed constraints.

Adlakha et al. [6] pointed out that as to apply the existing method [2], there is a need to find an optimal solution of the considered balanced fixed charge crisp transportation problem. Therefore, much computational efforts are required to apply the existing method [2]. To reduce the computational efforts, Adlakha et al. [6] proposed a basic feasible solution based heuristic algorithm to find a more-for-less solution of balanced fixed charge crisp transportation problems.

Storøy [20] proposed a method to find all more-for-less solutions of balanced crisp transportation problems.

Joshi and Gupta [15] proposed a heuristic algorithm to find an initial basic feasible solution of a special type of balanced crisp fractional transportation problems (balanced crisp fractional transportation problems in which the objective function is sum of linear function and linear fractional function). Joshi and Gupta [15] also proposed a condition under which at least one more-for-less solution will exist for the considered special type of balanced crisp fractional transportation problems. Furthermore, Joshi and Gupta [15] proposed a heuristic algorithm to find a more-for-less solution of the considered special type of balanced crisp fractional transportation problems.

Pandian and Natarajan [18] proposed a method to find an optimal solution of crisp transportation problems with mixed constraints. Pandian and Natarajan [18] also generalized the proposed method to find a more-for-less solution of crisp transportation problems with mixed constraints.

Basu et al. [8] proposed an algorithm to find all more-for-less solutions of balanced crisp transportation problems.

Joshi and Gupta [16] pointed out that as all existing conditions for the non-existence of more-for-less solution are dependent upon the dual variables of the linear fractional programming problem corresponding to the considered crisp linear fractional transportation problem. Therefore, much computational efforts are required to apply existing conditions for the non-existence of more-for-less solution. To reduce the computational efforts, Joshi and Gupta [16] proposed such a condition for the non-existence of more-for-less solution which is only dependent upon the cost matrix.

It is pertinent to mention that above-mentioned methods are proposed by considering the assumption that each parameter of a transportation problem is precisely known and hence each parameter can be represented by a positive real number. Therefore, above-mentioned methods cannot be used to find a more-for-less solution of such transportation problems with mixed constraints in which the precise value of a parameter is not known.

One of the ways, used in the literature to deal with imprecise parameters of transportation problems is to represent an imprecise parameter by a fuzzy set or any of its extensions [9]. One may also see Kaur et al. [17] and references therein for more details.

Pandian and Natarajan [19] proposed a method to find a more-for-less solution of such transportation problems with mixed constraints in which each imprecise parameter is represented by a non-negative triangular fuzzy number.

Hussain and Kumar [14] proposed a method to find a more-for-less solution of such transportation problems with mixed constraints in which the cost for supplying one unit quantity of the product from a source to a destination is represented by a positive real number. While, all other parameters (the availability of the product at each source and the demand at each destination) are represented by non-negative triangular intuitionistic fuzzy numbers.

Vidhya et al. [23] proposed a method to find a more-for-less solution of the existing fuzzy transportation problems with mixed constraints [19].

Remark It is pertinent to mention that in some works the term "more for less paradox" is used instead of "more-for-less solution".

1.3 Chapter-Wise Summary

The chapter-wise summary is as follows.

In Chap. 2, a new method (named as Mehar method-I) is proposed to check the existence of at least one more-for-less solution as well as to find all more-for-less solutions (if exists) of symmetric triangular fuzzy balanced transportation problems (balanced transportation problems in which each parameter and each variable is represented by a symmetric triangular fuzzy number). Also, to illustrate the proposed Mehar method-I, an existing crisp balanced transportation problem and a symmetric triangular fuzzy balanced transportation problem are solved by the proposed Mehar method-I.

In Chap. 3, a new method (named as Mehar method-II) is proposed to check the existence of at least one more-for-less solution as well as to find all more-for-less solutions (if exists) of symmetric triangular fuzzy transportation problems with mixed constraints (transportation problems with mixed constraints in which each parameter and each variable is represented by a symmetric triangular fuzzy number). Also, to illustrate the proposed Mehar method-II, an existing crisp transportation problem with mixed constraints and an existing symmetric triangular fuzzy transportation problem with mixed constraints are solved by the proposed Mehar method-II. Furthermore, the superiority of the proposed Mehar method-II over some existing methods is discussed.

In Chap. 4, firstly, the arithmetic operations of triangular fuzzy numbers as well as the method for comparing triangular fuzzy numbers, proposed by Vidhya et al. [23], are extended for triangular intuitionistic fuzzy numbers. Then, using the extended arithmetic operations and the extended comparing method, a new method (named as Mehar method-III) is proposed to find all more-for-less solutions of symmetric triangular intuitionistic fuzzy transportation problems with mixed constraints (transportation problems with mixed constraints in which each parameter and each variable is represented by a symmetric triangular intuitionistic fuzzy number). Finally, a symmetric triangular intuitionistic fuzzy transportation problem with mixed constraints is solved to illustrate the proposed Mehar method-III.

In Chap. 5, a new method (named as Mehar method-IV) is proposed to check the existence of at least one more-for-less solution as well as to find all more-for-less solutions (if exists) of symmetric triangular intuitionistic fuzzy linear fractional transportation problems with mixed constraints (linear fractional transportation problems with mixed constraints in which each parameter and each variable is represented by a symmetric triangular intuitionistic fuzzy number). Also, to illustrate the proposed Mehar method-IV, an existing crisp linear fractional transportation problem with mixed constraints and a symmetric triangular intuitionistic fuzzy linear fractional transportation problem with mixed constraints are solved by the proposed Mehar method-IV. Furthermore, the superiority of the proposed Mehar method-IV over an existing method is discussed.

In Chap. 6, some open research problems are suggested.

References

1. V. Adlakha, K. Kowalski, A quick sufficient solution to the more-for-less paradox in the transportation problem. Omega **26**, 541–547 (1998)
2. V. Adlakha, K. Kowalski, On the fixed-charge transportation problem. Omega **27**, 381–388 (1999)
3. V. Adlakha, K. Kowalski, A note on the procedure MFL for a more-for-less solution in transportation problems. Omega **28**, 481–483 (2000)
4. V. Adlakha, K. Kowalski, A heuristic method for 'more-for-less' in distribution-related problems. Int. J. Math. Educ. Sci. Technol. **32**, 61–71 (2001)
5. V. Adlakha, K. Kowalski, B. Lev, Solving transportation problems with mixed constraints. Int. J. Manage. Sci. Eng. Manage. **1**, 47–52 (2006)
6. V. Adlakha, K. Kowalski, R.R. Vemuganti, B. Lev, More-for-less algorithm for fixed-charge transportation problems. Omega **35**, 116–127 (2007)
7. H. Arsham, Postoptimality analyses of the transportation problem. J. Oper. Res. Soc. **43**, 121–139 (1992)
8. M. Basu, D. Acharya, A. Das, The algorithm of finding all paradoxical pairs in a linear transportation problem. Discr. Mathe. Algor. Appl. **4**, 1250049 (2012)
9. H. Bustince, E. Barrenechea, M. Pagola, J. Fernandez, Z. Xu, B. Bedregal, J. Montero, H. Hagras, F. Herrera, B.D. Baets, A historical account of types of fuzzy sets and their relationships. IEEE Trans. Fuzzy Syst. **24**, 179–194 (2016)
10. V.G. Deĭneko, B. Klinz, G.J. Woeginger, Which matrices are immune against the transportation paradox? Discr. Appl. Mathe. 130:495–501 (2003)
11. A. Gupta, S. Khanna, M.C. Puri, A paradox in linear fractional transportation problem with mixed constraints. Optimization **27**, 375–387 (1993)
12. A. Gupta, M.C. Puri, "More (same)-for-less" paradox in minimal cost network flow problem. Optimization **33**, 167–177 (1995)
13. F.L. Hitchcock, The distribution of a product from several sources to numerous localities. J. Math. Phys. **20**, 224–230 (1941)
14. R.J. Hussain, P.S. Kumar, An optimal more-for-less solution of mixed constraints intuitionistic fuzzy transportation problems. Int. J. Contemp. Math. Sci. **8**, 565–576 (2013)
15. V.D. Joshi, N. Gupta, On a paradox in linear plus linear fractional transportation problem. MATEMATIKA: Malaysian J. Ind. Appl. Math. **26**, 167–178 (2010)
16. V.D. Joshi, N. Gupta, Identifying more-for-less paradox in the linear fractional transportation problem using objective matrix. MATEMATIKA: Malaysian J. Ind. Appl. Math. **28**, 173–180 (2012)
17. A. Kaur, J. Kacprzyk, A. Kumar, *Fuzzy Transportation and Transshipment Problems, Studies in Fuzziness and Soft Computing* (Springer, Switzerland, 2020)

18. P. Pandian, G. Natarajan, A new method for finding an optimal more-for-less solution of transportation problems with mixed constraints. Int. J. Contemp. Math. Sci. **5**, 931–942 (2010)
19. P. Pandian, G. Natarajan, An optimal more-for-less solution to fuzzy transportation problems with mixed constraints. Appl. Math. Sci. **4**, 1405–1415 (2010)
20. S. Storøy, in The Transportation Paradox Revisited, N-5020 Bergen, Norway, 30 August 2007
21. W. Szwarc, The transportation paradox. Naval Res. Logist. Quart. **18**, 185–202 (1971)
22. V. Verma, M. C. Puri, in *On a Paradox in Linear Fractional Transportation Problem, Recent Developments in Mathematical Programming*, ed. by S. Kumar. Published on behalf of Australian Society for Operations Research (Gordan and Breach Science Publishers, 1991), pp. 413–424.
23. V. Vidhya, P. Uma Maheswari, K. Ganesan, An alternate method for finding more for less solution to fuzzy transportation problem with mixed constraints. Soft Comput. **25**, 11989–11996 (2021)

Chapter 2
Mehar Method-I to Find All More-For-Less Solutions of Symmetric Fuzzy Balanced Transportation Problems

In this chapter, a new method (named as Mehar method-I) is proposed to check the existence of at least one more-for-less solution as well as to find all more-for-less solutions (if exists) of symmetric triangular fuzzy balanced transportation problems (balanced transportation problems in which each parameter and each variable is represented by a symmetric triangular fuzzy number). Also, to illustrate the proposed Mehar method-I, an existing crisp balanced transportation problem and a symmetric triangular fuzzy balanced transportation problem are solved by the proposed Mehar method-I.

2.1 Some Basic Definitions

In this section, some basic definitions are discussed.

Definition 2.1.1 [3] A set $\tilde{A} = \{\langle x, \mu_{\tilde{A}}(x)\rangle : x \in X\}$, defined over the universal set X, is said to be a fuzzy set, where $\mu_{\tilde{A}}(x) \in [0, 1]$ represents the degree of membership for x belongs to the set $\tilde{A}$.

Definition 2.1.2 [3] Let $\tilde{A}$ be a fuzzy set defined over the universal set X and $\alpha \in (0, 1]$. Then, the crisp set $A_\alpha = \{x \in X : \mu_{\tilde{A}}(x) \geq \alpha\}$ is said to be the α-cut of the fuzzy set $\tilde{A}$.

Definition 2.1.3 [3] Let $\tilde{A}$ be a fuzzy set defined over the universal set X. Then, the crisp set $S\left(\tilde{A}\right) = \{x \in X : \mu_{\tilde{A}}(x) > 0\}$ is said to be the support of the fuzzy set $\tilde{A}$.

Definition 2.1.4 [3] Let $\tilde{A}$ be a fuzzy set defined over the universal set X. Then, the crisp number $h\left(\tilde{A}\right) = \underset{x \in X}{\text{supremum}}\left\{\mu_{\tilde{A}}(x)\right\}$ is said to be the height of the fuzzy set $\tilde{A}$. If $h\left(\tilde{A}\right) = 1$, then the fuzzy set $\tilde{A}$ is said to be a normal fuzzy set.

T. K. Bhatia et al., *More-for-Less Solutions in Fuzzy Transportation Problems*, Studies in Fuzziness and Soft Computing 426,
https://doi.org/10.1007/978-3-031-30337-1_2

Definition 2.1.5 [3] A fuzzy set $\tilde{A}$, defined over the set of real numbers, is said to be fuzzy number if it satisfies the following conditions.

(i) $\tilde{A}$ is normal,
(ii) A_α is a closed interval for every $\alpha \in (0, 1]$,
(iii) The support of $\tilde{A}$ is bounded.

Definition 2.1.6 [6] A fuzzy number $\tilde{A}$ is said to be triangular fuzzy number if its membership function is defined as.

$$\mu_{\tilde{A}}(x) = \begin{cases} \frac{x-a^1}{a^2-a^1}, a^1 < x \le a^2, \\ \frac{a^3-x}{a^3-a^2}, a^2 < x \le a^3, \\ 0, otherwise \end{cases}$$

A triangular fuzzy number $\tilde{A}$ may be represented as $\tilde{A} = \left(a^1, a^2, a^3\right)$ or $\tilde{A} = \left(a^2, a^2 - a^1, a^3 - a^2\right)$ or $\tilde{A} = \left(a^2, \beta(1-\alpha), \gamma(1-\alpha)\right)$ where $\alpha \in (0, 1]$, $\beta = a^2 - a^1$ and $\gamma = a^3 - a^2$.

Definition 2.1.7 [6] A triangular fuzzy number $\tilde{A} = \left(a^2, \beta(1-\alpha), \gamma(1-\alpha)\right)$, where $\alpha \in (0, 1]$ is said to be symmetric triangular fuzzy number if $\beta = \gamma$. Assuming $\beta = \gamma = \delta$, a symmetric triangular fuzzy number can be represented as $\tilde{A} = \left(a^2, \delta(1-\alpha), \delta(1-\alpha)\right)$, where $\alpha \in (0, 1]$.

Definition 2.1.8 [6] A symmetric triangular fuzzy number $\tilde{A} = \left(a^2, \delta(1-\alpha), \delta(1-\alpha)\right)$, where $\alpha \in (0, 1]$ is said to be zero triangular fuzzy number if $a^2 = 0$. Hence, a zero triangular fuzzy number can be represented as $\tilde{A} = (0, \delta(1-\alpha), \delta(1-\alpha))$, where δ is any non-negative real number and $\alpha \in (0, 1]$.

Definition 2.1.9 [6] Let $\tilde{A}_1 = \left(a_1^2, \beta_1(1-\alpha), \gamma_1(1-\alpha)\right)$ and $\tilde{A}_2 = \left(a_2^2, \beta_2(1-\alpha), \gamma_2(1-\alpha)\right)$, $\alpha \in (0, 1]$ be two triangular fuzzy numbers. Then,

(i) $\tilde{A}_1 \oplus \tilde{A}_2 = \begin{pmatrix} a_1^2 + a_2^2, maximum\{\beta_1(1-\alpha), \beta_2(1-\alpha)\}, \\ maximum\{\gamma_1(1-\alpha), \gamma_2(1-\alpha)\} \end{pmatrix}$

(ii) $\tilde{A}_1 \otimes \tilde{A}_2 = \begin{pmatrix} a_1^2 a_2^2, maximum\{\beta_1(1-\alpha), \beta_2(1-\alpha)\}, \\ maximum\{\gamma_1(1-\alpha), \gamma_2(1-\alpha)\} \end{pmatrix}$

Definition 2.1.10 [6] Let $\tilde{A}_1 = \left(a_1^2, \beta_1(1-\alpha), \gamma_1(1-\alpha)\right)$ and $\tilde{A}_2 = \left(a_2^2, \beta_2(1-\alpha), \gamma_2(1-\alpha)\right)$, $\alpha \in (0, 1]$ be two triangular fuzzy numbers. Then,

(i) $\tilde{A}_1 <_{Mag} \tilde{A}_2$ if $Mag\left(\tilde{A}_1\right) < Mag\left(\tilde{A}_2\right)$,
(ii) $\tilde{A}_1 \le_{Mag} \tilde{A}_2$ if $Mag\left(\tilde{A}_1\right) \le Mag\left(\tilde{A}_2\right)$,
(iii) $\tilde{A}_1 =_{Mag} \tilde{A}_2$ if $Mag\left(\tilde{A}_1\right) = Mag\left(\tilde{A}_2\right)$.

where,

$$Mag\left(\tilde{A}_i\right) = \frac{\gamma_i(1-\alpha) + 4a_i^2 - \beta_i(1-\alpha)}{4}; i = 1, 2 \tag{2.1}$$

Also, it can be easily verified that if $\tilde{A}_i = \left(a_i^2, \delta_i(1-\alpha), \delta_i(1-\alpha)\right); i = 1, 2, \ldots, k$ are symmetric triangular fuzzy numbers. Then,

(i)

$$Mag\left(\oplus_{i=1}^{k} \tilde{A}_i\right) = \sum_{i=1}^{k} Mag\left(\tilde{A}_i\right) \tag{2.2}$$

(ii)

$$Mag\left(\otimes_{i=1}^{k} \tilde{A}_i\right) = \prod_{i=1}^{k} Mag\left(\tilde{A}_i\right) \tag{2.3}$$

(iii) $Mag\left(\tilde{A}_i\right) = a_i^2$ (2.4)

2.2 Tabular Representation of Crisp Balanced Transportation Problems

A crisp balanced transportation problem can be represented into tabular form as shown by Table 2.1.

where,

(i) m represents the number of sources.
(ii) n represents the number of destinations.
(iii) S_i represents the ith source.

Table 2.1 Tabular representation of a crisp transportation problem

	D_1	D_2	...	D_n	Availability
S_1	c_{11}	c_{12}	...	c_{1n}	$= a_1$
S_2	c_{21}	c_{22}	...	c_{2n}	$= a_2$
⋮	⋮	⋮	⋮	⋮	⋮
S_m	c_{m1}	c_{m2}	...	c_{mn}	$= a_m$
Demand	$= b_1$	$= b_2$	...	$= b_n$	$\sum_{i=1}^{m} a_i = \sum_{j=1}^{n} b_j$

(iv) D_j represents the jth destination.
(v) The positive real number c_{ij} represents the cost for transporting one unit quantity of the product from the ith source to the jth destination.
(vi) The positive real number a_i represents the availability of the product at the ith source.
(vii) The positive real number b_j represents the demand of the product at the jth destination.

2.3 Tabular Representation of Symmetric Triangular Fuzzy Balanced Transportation Problems

A symmetric triangular fuzzy balanced transportation problem can be represented into tabular form as shown by Table 2.2.

where,

(i) The symmetric triangular fuzzy number $\tilde{c}_{ij} = \left(c_{ij}^2, \delta_{ij1}(1-\alpha), \delta_{ij1}(1-\alpha)\right)$ represents the fuzzy cost for transporting one unit quantity of product from the ith source to the jth destination.
(ii) The symmetric triangular fuzzy number $\tilde{a}_i = (a_i^2, \delta_{i2}(1-\alpha), \delta_{i2}(1-\alpha))$ represents the fuzzy supply of product at the ith source.
(iii) The symmetric triangular fuzzy number $\tilde{b}_j = \left(b_j^2, \delta_{j3}(1-\alpha), \delta_{j3}(1-\alpha)\right)$ represents the fuzzy demand of product at the jth destination.

Table 2.2 Tabular representation of a symmetric fuzzy balanced transportation problem

	D_1	D_2	…	D_n	Availability
S_1	$\tilde{c}_{11}$	$\tilde{c}_{12}$	…	$\tilde{c}_{1n}$	$=_{Mag}\tilde{a}_1$
S_2	$\tilde{c}_{21}$	$\tilde{c}_{22}$	…	$\tilde{c}_{2n}$	$=_{Mag}\tilde{a}_2$
⋮	⋮	⋮	⋮	⋮	⋮
S_m	$\tilde{c}_{m1}$	$\tilde{c}_{m2}$	…	$\tilde{c}_{mn}$	$=_{Mag}\tilde{a}_m$
Demand	$=_{Mag}\tilde{b}_1$	$=_{Mag}\tilde{b}_2$	…	$=_{Mag}\tilde{b}_n$	

2.4 Crisp Linear Programming Problems Corresponding to Crisp Balanced Transportation Problems

If it is assumed that the non-negative real number x_{ij} represents the quantity of the product to be supplied from the ith source to the jth destination and Z represents the total crisp transportation cost. Then, the crisp balanced transportation problem, represented by Table 2.1, can be formulated into its equivalent crisp linear programming problem ($P2.1$) or into its equivalent crisp linear programming problem ($P2.2$) (dual problem of the crisp linear programming problem ($P2.1$)).

Problem (***P***2.1) $Minimize\left(Z = \sum_{i=1}^{m}\sum_{j=1}^{n} c_{ij}x_{ij}\right)$
Subject to

$$\sum_{j=1}^{n} x_{ij} = a_i;\ i = 1, 2, \ldots, m;$$

$$\sum_{i=1}^{m} x_{ij} = b_j;\ j = 1, 2, \ldots, n;$$

$$x_{ij} \geq 0;\ i = 1, 2, \ldots, m;\ j = 1, 2, \ldots, n.$$

Problem (***P***2.2) $Maximize\left(Z = \sum_{i=1}^{m} a_i u_i + \sum_{j=1}^{n} b_j v_j\right)$
Subject to

$$u_i + v_j \leq c_{ij};\ i = 1, 2, \ldots, m;\ j = 1, 2, \ldots, n;$$

$$u_i, v_j \text{ are unrestricted variables; } i = 1, 2, \ldots, m;\ j = 1, 2, \ldots, n.$$

It is a well-known fact that the number of linearly independent constraints in the crisp linear programming problem ($P2.1$) will always be $(m + n - 1)$. If it is assumed that the first constraint of the crisp linear programming problem ($P2.1$) is depending upon the remaining $(m + n - 1)$ constraints. Then, the total optimal transportation cost of the crisp balanced transportation problem, represented by Table 2.1, can be either obtained by solving the crisp linear programming problem ($P2.3$) or by solving the crisp linear programming problem ($P2.4$) (dual of the crisp linear programming problem ($P2.3$)).

Problem (***P***2.3) $Minimize\left(Z = \sum_{i=1}^{m}\sum_{j=1}^{n} c_{ij}x_{ij}\right)$

Subject to

$$\sum_{j=1}^{n} x_{ij} = a_i;\ i = 2, \ldots, m;$$

$$\sum_{i=1}^{m} x_{ij} = b_j;\ j = 1, 2, \ldots, n;$$

$$x_{ij} \geq 0;\ i = 1, 2, \ldots, m;\ j = 1, 2, \ldots, n.$$

Problem ($\boldsymbol{P}$2.4) $Maximize\left(Z = \sum_{i=1}^{m} a_i u_i + \sum_{j=1}^{n} b_j v_j\right)$
Subject to

$$u_i + v_j \leq c_{ij};\ i = 1, 2, \ldots, m;\ j = 1, 2, \ldots, n;$$

$u_1 = 0,$
u_i, v_j are unrestricted variables; $i = 2, 3, \ldots, m;\ j = 1, 2, \ldots, n.$

2.5 Fuzzy Linear Programming Problems Corresponding to Symmetric Triangular Fuzzy Balanced Transportation Problems

If it is assumed that the symmetric triangular fuzzy number $\tilde{x}_{ij} = \left(x_{ij}^2, \delta_{ij4}(1-\alpha), \delta_{ij4}(1-\alpha)\right)$ represents the fuzzy quantity of the product to be supplied from the ith source to the jth destination, the symmetric triangular fuzzy number $\tilde{Z}$ represents the total fuzzy transportation cost and $\tilde{0}$ represents the zero triangular fuzzy number. Then, the symmetric triangular fuzzy balanced transportation problem, represented by Table 2.2, can be formulated into its equivalent fuzzy linear programming problem (P2.5) or into its equivalent crisp linear programming problem (P2.6) (dual problem of the fuzzy linear programming problem (P2.5)).

Problem ($\boldsymbol{P}$2.5) $Minimize\left(\tilde{Z} =_{Mag} \oplus_{i=1}^{m} \oplus_{j=1}^{n} (\tilde{c}_{ij} \otimes \tilde{x}_{ij})\right)$
Subject to
$\oplus_{j=1}^{n} \tilde{x}_{ij} =_{Mag} \tilde{a}_i;\ i = 1, 2, \ldots, m;$
$\oplus_{i=1}^{m} \tilde{x}_{ij} =_{Mag} \tilde{b}_j;\ j = 1, 2, \ldots, n;$
$\tilde{x}_{ij} \geq_{Mag} \tilde{0};\ i = 1, 2, \ldots, m;\ j = 1, 2, \ldots, n$

Problem (**P**2.6) $Maximize\left(Z^2 = \sum_{i=1}^{m} a_i^2 u_i^2 + \sum_{j=1}^{n} b_j^2 v_j^2\right)$
Subject to

$$u_i^2 + v_j^2 \le c_{ij}^2;\ i = 1, 2, \ldots, m;\ j = 1, 2, \ldots, n;$$

$$u_i^2, v_j^2 \text{ are unrestricted variables; } i = 1, 2, \ldots, m;\ j = 1, 2, \ldots, n.$$

It is pertinent to mention that the coefficient matrix corresponding to the constraints of the fuzzy linear programming problem (P2.5) is same as the coefficient matrix corresponding to the constraints of the crisp linear programming problem (P2.1). Therefore, the number of linearly independent constraints in the fuzzy linear programming problem (P2.5) will also be $(m + n - 1)$. If it is assumed that the first constraint of the fuzzy linear programming problem (P2.5) is depending upon the remaining $(m + n - 1)$ constraints. Then, the total optimal fuzzy transportation cost of the symmetric triangular fuzzy balanced transportation problem, represented by Table 2.2, can be either obtained by solving the fuzzy linear programming problem (P2.7) or by solving the crisp linear programming problem (P2.8) (dual of the fuzzy linear programming problem (P2.7)).

Problem (**P**2.7) $Minimize\left(\tilde{Z} =_{Mag} \oplus_{i=1}^{m} \oplus_{j=1}^{n} (\tilde{c}_{ij} \otimes \tilde{x}_{ij})\right)$
Subject to
$\oplus_{j=1}^{n} \tilde{x}_{ij} =_{Mag} \tilde{a}_i;\ i = 2, \ldots, m;$
$\oplus_{i=1}^{m} \tilde{x}_{ij} =_{Mag} \tilde{b}_j;\ j = 1, 2, \ldots, n;$
$\tilde{x}_{ij} \succeq_{Mag} \tilde{0};\ i = 1, 2, \ldots, m;\ j = 1, 2, \ldots, n$

Problem (**P**2.8) $Maximize\left(Z^2 = \sum_{i=2}^{m} a_i^2 u_i^2 + \sum_{j=1}^{n} b_j^2 v_j^2\right)$
Subject to

$$u_i^2 + v_j^2 \le c_{ij}^2;\ i = 1, 2, \ldots, m;\ j = 1, 2, \ldots, n;$$

$$u_1^2 = 0,$$

$$u_i^2, v_j^2 \text{ are unrestricted variables; } i = 2, 3, \ldots, m;\ j = 1, 2, \ldots, n.$$

The dual problem (P2.8) is obtained as follows.

Step 1: Using Definition 2.1.10, the fuzzy linear programming problem (P2.7) can be transformed into its equivalent crisp linear programming problem (P2.9).

Problem (**P 2.9** $Minimize\left(Mag\left(\tilde{Z}\right) = Mag\left(\oplus_{i=1}^{m} \oplus_{j=1}^{n} (\tilde{c}_{ij} \otimes \tilde{x}_{ij})\right)\right)$)
Subject to
$Mag\left(\oplus_{j=1}^{n} \tilde{x}_{ij}\right) = Mag(\tilde{a}_i);\ i = 2, \ldots, m$

$$Mag\left(\oplus_{i=1}^{m}\tilde{x}_{ij}\right) = Mag\left(\tilde{b}_j\right);\ j = 1, 2, \ldots, n$$

$$Mag\left(\tilde{x}_{ij}\right) \geq Mag\left(\tilde{0}\right);\ i = 1, 2, \ldots, m;\ j = 1, 2, \ldots, n.$$

Step 2: Using the expression (2.2), the crisp linear programming problem ($P2.9$) can be transformed into its equivalent crisp linear programming problem ($P2.10$).

Problem* (*P2.10 $Minimize\left(Mag\left(\tilde{Z}\right) = \left(\sum_{i=1}^{m}\sum_{j=1}^{n} Mag\left(\tilde{c}_{ij} \otimes \tilde{x}_{ij}\right)\right)\right)$)

Subject to

$\sum_{j=1}^{n} Mag\left(\tilde{x}_{ij}\right) = Mag(\tilde{a}_i);\ i = 2, \ldots, m;$

$\sum_{i=1}^{m} Mag\left(\tilde{x}_{ij}\right) = Mag\left(\tilde{b}_j\right);\ j = 1, 2, \ldots, n;$

$Mag\left(\tilde{x}_{ij}\right) \geq Mag\left(\tilde{0}\right);\ i = 1, 2, \ldots, m;\ j = 1, 2, \ldots, n.$

Step 3: Using the expression (2.3), the crisp linear programming problem ($P2.10$) can be transformed into its equivalent crisp linear programming problem ($P2.11$).

Problem (*P2.11*) $Minimize\left(Mag\left(\tilde{Z}\right) = \left(\sum_{i=1}^{m}\sum_{j=1}^{n} Mag\left(\tilde{c}_{ij}\right)Mag\left(\tilde{x}_{ij}\right)\right)\right)$

Subject to

Constraints of the problem ($P2.10$).

Step 4: Using the expression (2.4), the crisp linear programming problem ($P2.11$) can be transformed into its equivalent crisp linear programming problem ($P2.12$).

Problem (*P2.12*) $Minimize\left(Z^2 = \sum_{i=1}^{m}\sum_{j=1}^{n} c_{ij}^2 x_{ij}^2\right)$

Subject to

$\sum_{j=1}^{n} x_{ij}^2 = a_i^2;\ i = 2, \ldots, m$

$$\sum_{i=1}^{m} x_{ij}^2 = b_j^2;\ j = 1, 2, \ldots, n$$

$$x_{ij}^2 \geq 0;\ i = 1, 2, \ldots, m;\ j = 1, 2, \ldots, n.$$

Step 5: Using the duality theory [5], the crisp linear programming problem ($P2.8$) represents the dual of the crisp linear programming problem ($P2.12$). Hence, the crisp linear programming problem ($P2.8$) represents the dual of the fuzzy linear programming problem ($P2.7$).

Table 2.3 Tabular representation of a crisp balanced transportation problem

	D_1	D_2	...	D_n	Availability
S_1	c_{11}^2	c_{12}^2	...	c_{1n}^2	$= a_1^2$
S_2	c_{21}^2	c_{22}^2	...	c_{2n}^2	$= a_2^2$
$\vdots$	$\vdots$	$\vdots$	$\vdots$	$\vdots$	$\vdots$
S_m	c_{m1}^2	c_{m2}^2	...	c_{mn}^2	$= a_m^2$
Demand	$= b_1^2$	$= b_2^2$	...	$= b_n^2$	$\sum_{i=1}^{m} a_i^2 = \sum_{j=1}^{n} b_j^2$

2.6 Crisp Balanced Transportation Problems Equivalent to Symmetric Triangular Fuzzy Balanced Transportation Problems

The following clearly indicates that the symmetric triangular fuzzy balanced transportation problem, represented by Table 2.2, is equivalent to crisp balanced transportation problem represented by Table 2.3.

It is obvious from Step 1 to Step 4 of Sect. 2.5 that the fuzzy linear programming problem ($P2.3$) corresponding to the symmetric triangular balanced fuzzy transportation problem, represented by Table 2.2, is equivalent to the crisp linear programming problem ($P2.8$). Also, it is obvious from Sect. 2.4 that the crisp linear programming problem ($P2.8$) is equivalent to crisp balanced transportation problem represented by Table 2.3.

2.7 Proposed Sufficient Condition-I for the Existence of at Least One More-For-Less Solution

It is obvious from Sect. 2.6 that the symmetric triangular fuzzy balanced transportation problem, represented by Table 2.2, is equivalent to the crisp balanced transportation problem represented by Table 2.3. Hence, to propose a sufficient condition-I for the existence of at least one more-less-solution for the symmetric triangular fuzzy balanced transportation problem, represented by Table 2.2, is equivalent to propose a sufficient condition-I for the existence of at least one more-less-solution for the crisp balanced transportation problem represented by Table 2.3. Keeping the same in mind, in this section, the following existing sufficient condition [4] can be used to check the existence of at least one more-for-less solution of crisp balanced transportation problem represented by Table 2.3.

If in an optimal solution of the crisp linear programming problem ($P2.8$), there exist variables u_p^2 and v_q^2 such that $u_p^2 + v_q^2 < 0$. Then, at least one more-for-less solution of crisp balanced transportation problem, represented by Table 2.3, will exist. Hence, at least one more-for-less solution of symmetric triangular fuzzy balanced transportation problem, represented by Table 2.2, will exist.

This sufficient condition-I is obtained as follows.

According to the duality theory, to find the total minimum transportation cost for the crisp balanced transportation problem, represented by Table 2.3, is equivalent to find the optimal value of the objective function i.e., $Z^2 = \sum_{i=1}^{m}\sum_{j=1}^{n} c_{ij}^2 x_{ij}^2$ of the crisp linear programming problem ($P2.12$) or the optimal value of the objective function i.e., $Z^2 = \sum_{i=2}^{m} a_i^2 u_i^2 + \sum_{j=1}^{n} b_j^2 v_j^2$ of the dual problem ($P2.8$) of the crisp linear programming problem ($P2.12$).

Increasing the availability of the pth source and the demand of the qth destination by the positive quantity θ, the total crisp transportation cost of the modified problem is

$$Z_{modified}^2 = \sum_{\substack{i=2 \\ i \neq p}}^{m} a_i^2 u_i^2 + \left(a_p^2 + \theta\right) u_p^2 + \sum_{\substack{j=1 \\ j \neq q}}^{n} b_j^2 v_j^2 + \left(b_q^2 + \theta\right) v_q^2$$

$$Z_{modified}^2 = \sum_{\substack{i=2 \\ i \neq p}}^{m} a_i^2 u_i^2 + \left(a_p^2 u_p^2 + \theta u_p^2\right) + \sum_{\substack{j=1 \\ j \neq q}}^{n} b_j^2 v_j^2 + \left(b_q^2 v_q^2 + \theta v_q^2\right)$$

$$Z_{modified}^2 = \sum_{i=2}^{m} a_i^2 u_i^2 + \theta u_p^2 + \sum_{j=1}^{n} b_j^2 v_j^2 + \theta v_q^2 \tag{2.5}$$

Substituting $Z^2 = \sum_{i=2}^{m} a_i^2 u_i^2 + \sum_{j=1}^{n} b_j^2 v_j^2$ in (2.5),

$$Z_{modified}^2 = Z^2 + \theta u_p^2 + \theta v_q^2$$

Since, θ is a positive real number. So, the total crisp transportation cost i.e., $Z_{modified}^2 = Z^2 + \theta u_p^2 + \theta v_q^2$ of the modified problem will be less than the total crisp transportation cost i.e., Z^2 only if $u_p^2 + v_q^2 < 0$.

2.8 Proposed Mehar Method-I

In this section, a new method (named as Mehar method-I) is proposed to check the existence of at least one more-for-less solution as well as to find all more-for-less solutions (if exists) of symmetric triangular fuzzy balanced transportation problems.

The steps of the proposed Mehar method-I are as follows.

Step 1: Use **Step 1.1** to **Step 1.4** to check that for the symmetric triangular fuzzy balanced transportation problem, represented by Table 2.2, at least one more-for-less solution will exist or not.

Step 1.1: Transform the symmetric triangular fuzzy balanced transportation problem, represented by Table 2.2, into its equivalent crisp balanced transportation problem represented by Table 2.3.
Step 1.2: Find an optimal solution $u_1^2 = 0, u_i^2; i = 2, \ldots, m, v_j^2;\ j = 1, 2, \ldots, n$ of the problem ($P2.8$) by substituting the values of c_{ij}^2, a_i^2 and b_j^2 from Table 2.3.
Step 1.3: Using the optimal values of $u_i^2; i = 1, 2, \ldots, m, v_j^2; j = 1, 2, \ldots, n$, obtained in **Step 1.2**, construct a shadow price matrix $\left[A_{ij}\right]_{m \times n}$, where $A_{ij} = u_i^2 + v_j^2$.
Step 1.4: Check that all the elements A_{ij} of the shadow price matrix, obtained in **Step 1.3**, are non-negative or not.

Case (i): If all the elements of the shadow price matrix, obtained in **Step 1.3**, are non-negative. Then, there does not exist a more-for-less solution.
Case (ii): If at least one element of the shadow price matrix, obtained in **Step 1.3**, is negative. Then, at least one more-for-less solution will exist.

Step 2: If the element corresponding to the pth row and the qth column of the shadow price matrix, obtained in **Step 1.3**, is negative i.e., if $u_p^2 + v_q^2 < 0$. Then, a more-for-less solution can be obtained by increasing the availability of the pth source and the demand of the qth destination with the help of **Step 2.1** to **Step 2.8** of the post-optimality analysis [5].

Step 2.1: Construct the matrix

$$R_{pq} = \begin{bmatrix} a_2 \\ \vdots \\ a_p + \theta \\ \vdots \\ a_m \\ b_1 \\ b_2 \\ \vdots \\ b_q + \theta \\ \vdots \\ b_n \end{bmatrix}$$

where,

(i) θ is a positive real number.

(ii) a_i is the availability of the product at the ith source.

(iii) b_j is the demand of the product at the jth destination.

Step 2.2: Identify the basic variables in an optimal solution of the crisp linear programming problem ($P2.12$).

Step 2.3: Construct the basis matrix B with the help of the coefficients of the identified optimal basic variables in the constraints of the crisp linear programming problem ($P2.12$).

Step 2.4: Find the multiplicative inverse of the basis matrix B i.e., find B^{-1}.

Step 2.5: Find the matrix $B^{-1}R_{pq}$.

Step 2.6: Find the range of θ(where, $\theta > 0$) for which all the elements of the matrix $B^{-1}R_{pq}$, obtained in **Step 2.5**, are greater than or equal to zero.

Step 2.7: For any value of θ lying in its range, obtained in **Step 2.6**, the ith element (say, τ_i) of the matrix $B^{-1}R_{pq}$ represents the value of the ith basic variable in a more-for-less solution of crisp balanced transportation problem represented by Table 2.3.

Step 2.8: For any value of θ lying in its range, obtained in **Step 2.6**, the symmetric triangular fuzzy number $(\tau_i, \delta_i(1-\alpha), \delta_i(1-\alpha))$, where δ_i is any non-negative real number and $\alpha \in (0, 1]$, represents the value of the ith basic fuzzy variable in a more-for-less solution of the symmetric triangular fuzzy balanced transportation problem represented by Table 2.2.

2.9 Illustrative Examples

In this section, to illustrate the proposed Mehar method-I, all more-for-less solutions of an existing crisp balanced transportation problem [1, 2, 4] as well as a symmetric triangular fuzzy balanced transportation problem are obtained by the proposed Mehar method-I.

2.9.1 All More-For-Less Solutions of an Existing Problem

Using the proposed Mehar method-I, all more-for-less solutions (if exists) of the existing crisp balanced transportation problem [1, 2, 4], represented by Table 2.4, can be obtained as follows.

Step 1: According to **Step 1** of the proposed Mehar method-I, there is a need to use **Step 1.1** to **Step 1.4** to check that for the crisp balanced transportation problem, represented by Table 2.4, at least one more-for-less solution will exist or not.

Table 2.4 Crisp balanced transportation problem [1, 2, 4]

	D_1	D_2	D_3	D_4	D_5	Availability
S_1	14	15	6	13	14	7
S_2	16	9	22	13	16	18
S_3	8	5	11	4	5	6
S_4	12	4	18	9	10	15
Demand	4	11	12	8	11	

Step 1.1: According to **Step 1.1** of the proposed Mehar method-I, there is a need to transform the symmetric triangular fuzzy balanced transportation problem into its equivalent crisp balanced transportation problem. Since, the balanced transportation problem, represented by Table 2.4, is a crisp balanced transportation problem. So, there is no need to apply this step.

Step 1.2: According to **Step 1.2** of the proposed Mehar method-I, there is a need to find an optimal solution of the crisp linear programming problem (P2.13).

Problem (*P*2.13) *Maximize* $\begin{pmatrix} 7u_1^2 + 18u_2^2 + 6u_3^2 + 15u_4^2 + 4v_1^2 \\ +11v_2^2 + 12v_3^2 + 8v_4^2 + 11v_5^2 \end{pmatrix}$

Subject to

$$u_1^2 + v_1^2 \leq 14,$$

$$u_1^2 + v_2^2 \leq 15,$$

$$u_1^2 + v_3^2 \leq 6,$$

$$u_1^2 + v_4^2 \leq 13,$$

$$u_1^2 + v_5^2 \leq 14,$$

$$u_2^2 + v_1^2 \leq 16,$$

$$u_2^2 + v_2^2 \leq 9,$$

$$u_2^2 + v_3^2 \leq 22,$$

$$u_2^2 + v_4^2 \leq 13,$$

$$u_2^2 + v_5^2 \leq 16,$$

$$u_3^2 + v_1^2 \leq 8,$$

$$u_3^2 + v_2^2 \leq 5,$$

$$u_3^2 + v_3^2 \leq 11,$$

$$u_3^2 + v_4^2 \leq 4,$$

$$u_3^2 + v_5^2 \leq 5,$$

$$u_4^2 + v_1^2 \leq 12,$$

$$u_4^2 + v_2^2 \leq 4,$$

$$u_4^2 + v_3^2 \leq 18,$$

$$u_4^2 + v_4^2 \leq 9,$$

$$u_4^2 + v_5^2 \leq 10,$$

$$u_1^2 = 0,$$

$$u_2^2, u_3^2, v_1^2, v_2^2, v_3^2, \text{ are unrestricted in sign.}$$

It can be easily verified that on solving the problem ($P2.13$), the obtained optimal solution is $u_1^2 = 0, u_2^2 = 15, u_3^2 = 5, u_4^2 = 10, v_1^2 = 1, v_2^2 = -6, v_3^2 = 6, v_4^2 = -2, v_5^2 = 0$.

Step 1.3: According to **Step 1.3** of the proposed Mehar method-I, the following shadow price matrix is obtained.

$$\begin{bmatrix} u_1^2 + v_1^2 = 1 & u_1^2 + v_2^2 = -6 & u_1^2 + v_3^2 = 6 & u_1^2 + v_4^2 = -2 & u_1^2 + v_5^2 = 0 \\ u_2^2 + v_1^2 = 16 & u_2^2 + v_2^2 = 9 & u_2^2 + v_3^2 = 21 & u_2^2 + v_4^2 = 13 & u_2^2 + v_5^2 = 15 \\ u_3^2 + v_1^2 = 6 & u_3^2 + v_2^2 = -1 & u_3^2 + v_3^2 = 11 & u_3^2 + v_4^2 = 3 & u_3^2 + v_5^2 = 5 \\ u_4^2 + v_1^2 = 11 & u_4^2 + v_2^2 = 4 & u_4^2 + v_3^2 = 16 & u_4^2 + v_4^2 = 8 & u_4^2 + v_5^2 = 10 \end{bmatrix}$$

Step 1.4: Since, some elements of the shadow price matrix, obtained in **Step 1.3**, are negative. So, according to Case (ii) of **Step 1.4** of the proposed Mehar method-I, there will exist at least one more-for-less solution of the crisp balanced transportation problem represented by Table 2.4

Step 2: According to Step 2 of the proposed Mehar method-I, all more-for-less solutions of the crisp balanced transportation problem, represented by Table 2.4, can be obtained as follows.

Step 2.1: It is obvious from **Step 1.3** that $u_1^2 + v_2^2 < 0, u_1^2 + v_4^2 < 0$ and $u_3^2 + v_2^2 < 0$. Therefore, according to **Step 2.1** of the proposed Mehar method-I,

(i) $$R_{12} = \begin{bmatrix} 18 \\ 6 \\ 15 \\ 4 \\ 11+\theta \\ 12 \\ 8 \\ 11 \end{bmatrix}$$

(ii) $$R_{14} = \begin{bmatrix} 18 \\ 6 \\ 15 \\ 4 \\ 11 \\ 12 \\ 8+\theta \\ 11 \end{bmatrix}$$

(iii) $$R_{32} = \begin{bmatrix} 18 \\ 6+\theta \\ 15 \\ 4 \\ 11+\theta \\ 12 \\ 8 \\ 11 \end{bmatrix}$$

Step 2.2: According to **Step 2.2** of the proposed Mehar method-I, there is a need to identify the basic variables in an optimal solution of the crisp linear programming problem ($P2.14$).

Problem (*P*2.14) *Minimize* $\begin{pmatrix} Z = 14x_{11}^2 + 15x_{12}^2 + 6x_{13}^2 + 13x_{14}^2 \\ +14x_{15}^2 + 16x_{21}^2 + 9x_{22}^2 + 22x_{23}^2 \\ +13x_{24}^2 + 16x_{25}^2 + 8x_{31}^2 + 5x_{32}^2 \\ +11x_{33}^2 + 4x_{34}^2 + 5x_{35}^2 + 12x_{41}^2 \\ +4x_{42}^2 + 18x_{43}^2 + 9x_{44}^2 + 10x_{45}^2 \end{pmatrix}$

Subject to

$$x_{21}^2 + x_{22}^2 + x_{23}^2 + x_{24}^2 + x_{25}^2 = 18,$$

$$x_{31}^2 + x_{32}^2 + x_{33}^2 + x_{34}^2 + x_{35}^2 = 6,$$

$$x_{41}^2 + x_{42}^2 + x_{43}^2 + x_{44}^2 + x_{45}^2 = 15,$$

$$x_{11}^2 + x_{21}^2 + x_{31}^2 + x_{41}^2 = 4,$$

$$x_{12}^2 + x_{22}^2 + x_{32}^2 + x_{42}^2 = 11,$$

$$x_{13}^2 + x_{23}^2 + x_{33}^2 + x_{43}^2 = 12,$$

$$x_{14}^2 + x_{24}^2 + x_{34}^2 + x_{44}^2 = 8,$$

$$x_{15}^2 + x_{25}^2 + x_{35}^2 + x_{45}^2 = 11,$$

$$x_{ij}^2 \geq 0; i = 1, 2, 3, 4; j = 1, 2, 3, 4, 5.$$

It can be easily verified that the following optimal solution is obtained on solving the crisp linear programming problem (*P*2.14).

$x_{11}^2 = 0, x_{12}^2 = 0, x_{13}^2 = 7, x_{14}^2 = 0, x_{15}^2 = 0, x_{21}^2 = 4, x_{22}^2 = 6, x_{23}^2 = 0, x_{24}^2 = 8, x_{25}^2 = 0, x_{31}^2 = 0, x_{32}^2 = 0, x_{33}^2 = 5, x_{34}^2 = 0, x_{35}^2 = 1, x_{41}^2 = 0, x_{42}^2 = 5, x_{43}^2 = 0, x_{44}^2 = 0, x_{45}^2 = 10.$

Also, it can be easily verified that the basic variables in this obtained optimal solution are $x_{13}^2, x_{21}^2, x_{22}^2, x_{24}^2, x_{33}^2, x_{35}^2, x_{42}^2$ and x_{45}^2.

Step 2.3: According to **Step 2.3** of the proposed Mehar method-I, there is a need to construct the basis matrix B with the help of the coefficients of optimal basic variables $x_{13}^2, x_{21}^2, x_{22}^2, x_{24}^2, x_{33}^2, x_{35}^2, x_{42}^2, x_{45}^2$ in the constraints of the crisp linear programming problem (*P*2.14).

$$B = \begin{bmatrix} 0\,1\,1\,1\,0\,0\,0\,0 \\ 0\,0\,0\,0\,1\,1\,0\,0 \\ 0\,0\,0\,0\,0\,0\,1\,1 \\ 0\,1\,0\,0\,0\,0\,0\,0 \\ 0\,0\,1\,0\,0\,0\,1\,0 \\ 1\,0\,0\,0\,1\,0\,0\,0 \\ 0\,0\,0\,1\,0\,0\,0\,0 \\ 0\,0\,0\,0\,0\,1\,0\,1 \end{bmatrix}$$

Step 2.4: According to **Step 2.4** of the proposed Mehar method-I, there is a need to find the multiplicative inverse of the basis matrix B.

$$B^{-1} = \begin{bmatrix} -1 & -1 & -1 & 1 & 1 & 1 & 1 & 1 \\ 0 & 0 & 0 & 1 & 0 & 0 & 0 & 0 \\ 1 & 0 & 0 & -1 & 0 & 0 & -1 & 0 \\ 0 & 0 & 0 & 0 & 0 & 0 & 1 & 0 \\ 1 & 1 & 1 & -1 & -1 & 0 & -1 & -1 \\ -1 & 0 & -1 & 1 & 1 & 0 & 1 & 1 \\ -1 & 0 & 0 & 1 & 1 & 0 & 1 & 0 \\ 1 & 0 & 1 & -1 & -1 & 0 & -1 & 0 \end{bmatrix}$$

Step 2.5: According to **Step 2.5** of the proposed Mehar method-I, there is a need to find $B^{-1}R_{12}$, $B^{-1}R_{14}$ and $B^{-1}R_{32}$

.

(i) $$B^{-1}R_{12} = \begin{bmatrix} -1 & -1 & -1 & 1 & 1 & 1 & 1 & 1 \\ 0 & 0 & 0 & 1 & 0 & 0 & 0 & 0 \\ 1 & 0 & 0 & -1 & 0 & 0 & -1 & 0 \\ 0 & 0 & 0 & 0 & 0 & 0 & 1 & 0 \\ 1 & 1 & 1 & -1 & -1 & 0 & -1 & -1 \\ -1 & 0 & -1 & 1 & 1 & 0 & 1 & 1 \\ -1 & 0 & 0 & 1 & 1 & 0 & 1 & 0 \\ 1 & 0 & 1 & -1 & -1 & 0 & -1 & 0 \end{bmatrix} \begin{bmatrix} 18 \\ 6 \\ 15 \\ 4 \\ 11+\theta \\ 12 \\ 8 \\ 11 \end{bmatrix} = \begin{bmatrix} 7+\theta \\ 4+0\theta \\ 6+0\theta \\ 8+0\theta \\ 5-\theta \\ 1+\theta \\ 5+\theta \\ 10-\theta \end{bmatrix}$$

(ii) $$B^{-1}R_{14} = \begin{bmatrix} -1 & -1 & -1 & 1 & 1 & 1 & 1 & 1 \\ 0 & 0 & 0 & 1 & 0 & 0 & 0 & 0 \\ 1 & 0 & 0 & -1 & 0 & 0 & -1 & 0 \\ 0 & 0 & 0 & 0 & 0 & 0 & 1 & 0 \\ 1 & 1 & 1 & -1 & -1 & 0 & -1 & -1 \\ -1 & 0 & -1 & 1 & 1 & 0 & 1 & 1 \\ -1 & 0 & 0 & 1 & 1 & 0 & 1 & 0 \\ 1 & 0 & 1 & -1 & -1 & 0 & -1 & 0 \end{bmatrix} \begin{bmatrix} 18 \\ 6 \\ 15 \\ 4 \\ 11 \\ 12 \\ 8+\theta \\ 11 \end{bmatrix} = \begin{bmatrix} 7+\theta \\ 4+0\theta \\ 6-\theta \\ 8+\theta \\ 5-\theta \\ 1+\theta \\ 5+\theta \\ 10-\theta \end{bmatrix}$$

$$\text{(iii)}\quad B^{-1}R_{32} = \begin{bmatrix} -1 & -1 & -1 & 1 & 1 & 1 & 1 & 1 \\ 0 & 0 & 0 & 1 & 0 & 0 & 0 & 0 \\ 1 & 0 & 0 & -1 & 0 & 0 & -1 & 0 \\ 0 & 0 & 0 & 0 & 0 & 0 & 1 & 0 \\ 1 & 1 & 1 & -1 & -1 & 0 & -1 & -1 \\ -1 & 0 & -1 & 1 & 1 & 0 & 1 & 1 \\ -1 & 0 & 0 & 1 & 1 & 0 & 1 & 0 \\ 1 & 0 & 1 & -1 & -1 & 0 & -1 & 0 \end{bmatrix} \begin{bmatrix} 18 \\ 6+\theta \\ 15 \\ 4 \\ 11+\theta \\ 12 \\ 8 \\ 11 \end{bmatrix} = \begin{bmatrix} 7+0\theta \\ 4+0\theta \\ 6+0\theta \\ 8+0\theta \\ 5+0\theta \\ 1+\theta \\ 5+\theta \\ 10-\theta \end{bmatrix}$$

Step 2.6: According to **Step 2.6** of the proposed Mehar method-I, there is a need to find the range of θ (where, $\theta > 0$) so that all the elements of the matrix, obtained in **Step 2.5**, are greater than or equal to zero.

It is obvious from **Step 2.5** that

(i) The range of θ for which all the elements of the matrix $B^{-1}R_{12}$, obtained in **Step 2.5**, are greater than or equal to zero, i.e., $7+\theta \geq 0, 5-\theta \geq 0, 1+\theta \geq 0, 5+\theta \geq 0, 10-\theta \geq 0$ is $0 < \theta \leq 5$.

(ii) The range of θ for which all the elements of the matrix $B^{-1}R_{14}$, obtained in **Step 2.5**, are greater than or equal to zero, i.e., $7+\theta \geq 0, 6-\theta \geq 0, 8+\theta \geq 0, 5-\theta \geq 0, 1+\theta \geq 0, 5+\theta \geq 0, 10-\theta \geq 0$ is $0 < \theta \leq 5$.

(iii) The range of θ for which all the elements of the matrix $B^{-1}R_{32}$, obtained in **Step 2.5**, are greater than or equal to zero, i.e., $1+\theta \geq 0, 5+\theta \geq 0, 10-\theta \geq 0$ is $0 < \theta \leq 10$.

Step 2.7: According to **Step 2.7** of the proposed Mehar method-I,

(i) For any value θ such that $0 < \theta \leq 5$,

$7+\theta$ represents the value of the 1st basic variable i.e., x_{13}^2,
$4+0\theta$ represents the value of the 2nd basic variable i.e., x_{21}^2,
$6+0\theta$ represents the value of the 3rd variable i.e., x_{22}^2,
$8+0\theta$ represents the value of the 4th basic variable i.e., x_{24}^2,
$5-\theta$ represents the value of the 5th basic variable i.e., x_{33}^2,
$1+\theta$ represents the value of the 5th basic variable i.e., x_{35}^2,
$5+\theta$ represents the value of the 5th basic variable i.e., x_{42}^2, and.
$10-\theta$ represents the value of the 6th variable i.e., x_{45}^2 in a more-for-less solution of crisp balanced transportation problem represented by Table 2.4.

For example, substituting $\theta = 5$, the obtained values of basic variables in a more-for-less solution are

$$x_{13}^2 = 12,$$

$$x_{21}^2 = 4,$$

$$x_{22}^2 = 6,$$

$$x_{24}^2 = 8,$$

$$x_{33}^2 = 0,$$

$$x_{35}^2 = 6,$$

$$x_{42}^2 = 10,$$

$$x_{45}^2 = 5.$$

(ii) For any value θ such that $0 < \theta \leq 5$,
$7 + \theta$ represents the value of the 1st basic variable i.e., x_{13}^2,
$4 + 0\theta$ represents the value of the 2nd basic variable i.e., x_{21}^2,
$6 - \theta$ represents the value of the 3rd variable i.e., x_{22}^2,
$8 + \theta$ represents the value of the 4th basic variable i.e., x_{24}^2,
$5 - \theta$ represents the value of the 5th basic variable i.e., x_{33}^2,
$1 + \theta$ represents the value of the 5th basic variable i.e., x_{35}^2,
$5 + \theta$ represents the value of the 5th basic variable i.e., x_{42}^2, and.
$10 - \theta$ represents the value of the 6th variable i.e., x_{45}^2 in a more-for-less solution of crisp balanced transportation problem represented by Table 2.4.

For example, substituting $\theta = 5$, the obtained values of basic variables in a more-for-less solution are

$$x_{13}^2 = 12,$$

$$x_{21}^2 = 4,$$

$$x_{22}^2 = 1,$$

$$x_{24}^2 = 13,$$

$$x_{33}^2 = 0,$$

$$x_{35}^2 = 6,$$

$$x_{42}^2 = 10,$$

$$x_{45}^2 = 5.$$

(iii) For any value θ such that $0 < \theta \leq 10$,
$7 + 0\theta$ represents the value of the 1st basic variable i.e., x_{13}^2,
$4 + 0\theta$ represents the value of the 2nd basic variable i.e., x_{21}^2,
$6 + 0\theta$ represents the value of the 3rd variable i.e., x_{22}^2,
$8 + 0\theta$ represents the value of the 4th basic variable i.e., x_{24}^2,
$5 + 0\theta$ represents the value of the 5th basic variable i.e., x_{33}^2,
$1 + \theta$ represents the value of the 5th basic variable i.e., x_{35}^2,
$5 + \theta$ represents the value of the 5th basic variable i.e., x_{42}^2, and.
$10 - \theta$ represents the value of the 6th variable i.e., x_{45}^2 in a more-for-less solution of crisp balanced transportation problem represented by Table 2.4.

For example, substituting $\theta = 10$, the obtained values of basic variables in a more-for-less solution are

$$x_{13}^2 = 7,$$

$$x_{21}^2 = 4,$$

$$x_{22}^2 = 6,$$

$$x_{24}^2 = 8,$$

$$x_{33}^2 = 5,$$

$$x_{35}^2 = 11,$$

$$x_{42}^2 = 15,$$

$$x_{45}^2 = 0.$$

2.9.2 All More-For-Less Solutions of Considered Problem

Using the proposed Mehar method-I, all more-for-less solutions (if exists) of the symmetric triangular fuzzy balanced transportation problem, represented by Table 2.5, can be obtained as follows.

where,

(i) $\tilde{c}_{11} = (1, 1(1-\alpha), 1(1-\alpha)), \tilde{c}_{12} = (6, 3(1-\alpha), 3(1-\alpha)), \tilde{c}_{13} = (3, 2(1-\alpha), 2(1-\alpha)), \tilde{c}_{14} = (5, 3(1-\alpha), 3(1-\alpha)), \tilde{c}_{21} =$

Table 2.5 Symmetric triangular fuzzy balanced transportation problem

	D_1	D_2	D_3	D_4	Availability
S_1	$\tilde{c}_{11}$	$\tilde{c}_{12}$	$\tilde{c}_{13}$	$\tilde{c}_{14}$	$=_{Mag}\tilde{a}_1$
S_2	$\tilde{c}_{21}$	$\tilde{c}_{22}$	$\tilde{c}_{23}$	$\tilde{c}_{24}$	$=_{Mag}\tilde{a}_2$
S_3	$\tilde{c}_{31}$	$\tilde{c}_{32}$	$\tilde{c}_{33}$	$\tilde{c}_{34}$	$=_{Mag}\tilde{a}_3$
Demand	$=_{Mag}\tilde{b}_1$	$=_{Mag}\tilde{b}_2$	$=_{Mag}\tilde{b}_3$	$=_{Mag}\tilde{b}_4$	

$(7, 3(1-\alpha), 3(1-\alpha)), \tilde{c}_{22} = (3, 1(1-\alpha), 1(1-\alpha)), \tilde{c}_{23} = (1, 1(1-\alpha), 1(1-\alpha)), \tilde{c}_{24} = (6, 2(1-\alpha), 2(1-\alpha)), \tilde{c}_{31} = (9, 5(1-\alpha), 5(1-\alpha)), \tilde{c}_{32} = (4, 2(1-\alpha), 2(1-\alpha)), \tilde{c}_{33} = (5, 4(1-\alpha), 4(1-\alpha)), \tilde{c}_{34} = (4, 1(1-\alpha), 1(1-\alpha))$

(ii) $\tilde{a}_1 = (20, 8(1-\alpha), 8(1-\alpha)), \tilde{a}_2 = (10, 3(1-\alpha), 3(1-\alpha)), \tilde{a}_3 = (25, 10(1-\alpha), 10(1-\alpha))$

(iii) $\tilde{b}_1 = (11, 6(1-\alpha), 6(1-\alpha)), \tilde{b}_2 = (13, 4(1-\alpha), 4(1-\alpha)), \tilde{b}_3 = (17, 5(1-\alpha), 5(1-\alpha)), \tilde{b}_4 = (14, 7(1-\alpha), 7(1-\alpha))$

Step 1: According to **Step 1** of the proposed Mehar method-I, there is a need to use **Step 1.1** to **Step 1.4** to check that for the symmetric triangular fuzzy balanced transportation problem, represented by Table 2.5, at least one more-for-less solution will exist or not.

Step 1.1: According to **Step 1.1** of the proposed Mehar method-I, the symmetric triangular fuzzy balanced transportation problem, represented by Table 2.5, can be transformed into its equivalent crisp balanced transportation problem represented by Table 2.6.

Step 1.2: According to **Step 1.2** of the proposed Mehar method-I, there is a need to find an optimal solution of the crisp linear programming problem (P2.15).

Problem (P 2.15) $Maximize(20u_1^2 + 10u_2^2 + 25u_3^2 + 11v_1^2 + 13v_2^2 + 17v_3^2 + 14v_4^2)$

Subject to

$$u_1^2 + v_1^2 \le 1,$$

$$u_1^2 + v_2^2 \le 6,$$

$$u_1^2 + v_3^2 \le 3$$

$$u_1^2 + v_4^2 \le 5$$

$$u_2^2 + v_1^2 \le 7,$$

$$u_2^2 + v_2^2 \le 3,$$

$$u_2^2 + v_3^2 \le 1,$$

$$u_2^2 + v_4^2 \le 6,$$

Table 2.6 Crisp balanced transportation problem

	D_1	D_2	D_3	D_4	Availability
S_1	1	6	3	5	20
S_2	7	3	1	6	10
S_3	9	4	5	4	25
Demand	11	13	17	14	

$$u_3^2 + v_1^2 \leq 9,$$

$$u_3^2 + v_2^2 \leq 4,$$

$$u_3^2 + v_3^2 \leq 5,$$

$$u_3^2 + v_4^2 \leq 4,$$

$$u_1^2 = 0,$$

$$u_2^2, u_3^2, v_1^2, v_2^2, v_3^2 \text{ are unrestricted in sign.}$$

It can be easily verified that on solving the problem ($P2.15$), the obtained optimal solution is $u_1^2 = 0, u_2^2 = -2, u_3^2 = -1, v_1^2 = 1, v_2^2 = 5, v_3^2 = 3, v_4^2 = 5$.

Step 1.3: According to **Step 1.3** of the proposed Mehar method-I, the following shadow price matrix is obtained.

$$\begin{bmatrix} u_1^2 + v_1^2 = 1 & u_1^2 + v_2^2 = 5 & u_1^2 + v_3^2 = 3 & u_1^2 + v_4^2 = 5 \\ u_2^2 + v_1^2 = -1 & u_2^2 + v_2^2 = 3 & u_2^2 + v_3^2 = 1 & u_2^2 + v_4^2 = 3 \\ u_3^2 + v_1^2 = 0 & u_3^2 + v_2^2 = 4 & u_3^2 + v_3^2 = 2 & u_3^2 + v_4^2 = 4 \end{bmatrix}$$

Step 1.4: Since, one element of the shadow price matrix, obtained in **Step 1.3**, is negative. So, according to Case (ii) of **Step 1.4** of the proposed Mehar method-I, there will exist at least one more-for-less solution of symmetric triangular fuzzy balanced transportation problem represented by Table 2.5.

Step 2: According to Step 2 of the proposed Mehar method-I, all more-for-less solutions of the symmetric triangular fuzzy balanced transportation problem, represented by Table 2.5, can be obtained as follows.

Step 2.1: It is obvious from **Step 1.3** that $u_2^2 + v_1^2 < 0$. Therefore, according to **Step 2.1** of the proposed Mehar method-I,

$$R_{21} = \begin{bmatrix} 10 + \theta \\ 25 \\ 11 + \theta \\ 13 \\ 17 \\ 14 \end{bmatrix}$$

Step 2.2: According to **Step 2.2** of the proposed Mehar method-I, there is a need to identify the basic variables in an optimal solution of the crisp linear programming problem ($P2.16$).

Problem (*P*2.16) *Minimize* $\begin{pmatrix} Z = x_{11}^2 + 6x_{12}^2 + 3x_{13}^2 \\ +5x_{14}^2 + 7x_{21}^2 + 3x_{22}^2 \\ +x_{23}^2 + 6x_{24}^2 + 9x_{31}^2 \\ +4x_{32}^2 + 5x_{33}^2 + 4x_{34}^2 \end{pmatrix}$

Subject to

$$x_{21}^2 + x_{22}^2 + x_{23}^2 + x_{24}^2 = 10,$$

$$x_{31}^2 + x_{32}^2 + x_{33}^2 + x_{34}^2 = 25,$$

$$x_{11}^2 + x_{21}^2 + x_{31}^2 = 11,$$

$$x_{12}^2 + x_{22}^2 + x_{32}^2 = 13,$$

$$x_{13}^2 + x_{23}^2 + x_{33}^2 = 17,$$

$$x_{14}^2 + x_{24}^2 + x_{34}^2 = 14,$$

$$x_{ij}^2 \geq 0; i = 1, 2, 3; j = 1, 2, 3, 4.$$

It can be easily verified that the following optimal solution is obtained on solving the crisp linear programming problem ($P2.16$).

$x_{11}^2 = 11, x_{12}^2 = 0, x_{13}^2 = 9, x_{14}^2 = 0, x_{21}^2 = 0, x_{22}^2 = 2, x_{23}^2 = 8, x_{24}^2 = 0, x_{31}^2 = 0, x_{32}^2 = 11, x_{33}^2 = 0, x_{34}^2 = 14.$

Also, it can be easily verified that the basic variables in this obtained optimal solution are $x_{11}^2, x_{13}^2, x_{22}^2, x_{23}^2, x_{32}^2$ and x_{34}^2.

Step 2.3: According to **Step 2.3** of the proposed Mehar method-I, there is a need to construct the basis matrix B with the help of the coefficients of optimal basic variables $x_{11}^2, x_{13}^2, x_{22}^2, x_{23}^2, x_{32}^2, x_{34}^2$ in the constraints of the crisp linear programming problem ($P2.16$).

$$B = \begin{bmatrix} 0\,0\,1\,1\,0\,0 \\ 0\,0\,0\,0\,1\,1 \\ 1\,0\,0\,0\,0\,0 \\ 0\,0\,1\,0\,1\,0 \\ 0\,1\,0\,1\,0\,0 \\ 0\,0\,0\,0\,0\,1 \end{bmatrix}$$

Step 2.4: According to **Step 2.4** of the proposed Mehar method-I, there is a need to find the multiplicative inverse of the basis matrix B.

$$B^{-1} = \begin{bmatrix} 0 & 0 & 1 & 0 & 0 & 0 \\ -1 & -1 & 0 & 1 & 1 & 1 \\ 0 & -1 & 0 & 1 & 0 & 1 \\ 1 & 1 & 0 & -1 & 0 & -1 \\ 0 & 1 & 0 & 0 & 0 & -1 \\ 0 & 0 & 0 & 0 & 0 & 1 \end{bmatrix}$$

Step 2.5: According to **Step 2.5** of the proposed Mehar method-I, there is need to find $B^{-1}R_{21}$.

$$B^{-1}R_{21} = \begin{bmatrix} 0 & 0 & 1 & 0 & 0 & 0 \\ -1 & -1 & 0 & 1 & 1 & 1 \\ 0 & -1 & 0 & 1 & 0 & 1 \\ 1 & 1 & 0 & -1 & 0 & -1 \\ 0 & 1 & 0 & 0 & 0 & -1 \\ 0 & 0 & 0 & 0 & 0 & 1 \end{bmatrix} \begin{bmatrix} 10+\theta \\ 25 \\ 11+\theta \\ 13 \\ 17 \\ 14 \end{bmatrix} = \begin{bmatrix} 11+\theta \\ 9-\theta \\ 2+0\theta \\ 8+\theta \\ 11+0\theta \\ 14+0\theta \end{bmatrix}$$

Step 2.6: According to **Step 2.6** of the proposed Mehar method-I, there is a need to find the range of θ (where, $\theta > 0$) so that all the elements of the matrix, obtained in **Step 2.5**, are greater than or equal to zero.

It is obvious from **Step 2.5** that the range of θ for which all the elements of the matrix $B^{-1}R_{21}$, obtained in **Step 2.5**, are greater than or equal to zero, i.e., $11+\theta \geq 0, 9-\theta \geq 0, 8+\theta \geq 0$ is $0 < \theta \leq 9$.

Step 2.7: According to **Step 2.7** of the proposed Mehar method-I, for any value θ such that $0 < \theta \leq 9$,

$11+\theta$ represents the value of the 1st basic variable i.e., x^2_{11},
$9-\theta$ represents the value of the 2nd basic variable i.e., x^2_{13},
$2+0\theta$ represents the value of the 3rd variable i.e., x^2_{22},
$8+\theta$ represents the value of the 4th basic variable i.e., x^2_{23},
$11+0\theta$ represents the value of the 5th basic variable i.e., x^2_{32} and.
$14+0\theta$ represents the value of the 6th variable i.e., x^2_{34} in a more-for-less solution of crisp balanced transportation problem represented by Table 2.6.

For example, substituting $\theta = 9$, the obtained values of basic variables in a more-for-less solution are

$$x^2_{11} = 20,$$

$$x^2_{13} = 0,$$

$$x^2_{22} = 2,$$

$$x_{23}^2 = 17,$$

$$x_{32}^2 = 11,$$

$$x_{34}^2 = 14$$

Step 2.8: According to **Step 2.8** of the proposed Mehar method-I, for any value θ such that $0 < \theta \leq 9$,

$(11 + \theta, \delta_{114}(1 - \alpha), \delta_{114}(1 - \alpha))$ represents the value of the 1st basic variable i.e., $\tilde{x}_{11}$,

$(9 - \theta, \delta_{134}(1 - \alpha), \delta_{134}(1 - \alpha))$ represents the value of the 2nd basic variable i.e., $\tilde{x}_{13}$,

$(2 + 0\theta, \delta_{224}(1 - \alpha), \delta_{224}(1 - \alpha))$ represents the value of the 3rd variable i.e., $\tilde{x}_{22}$,

$(8 + \theta, \delta_{234}(1 - \alpha), \delta_{234}(1 - \alpha))$ represents the value of the 4th basic variable i.e., $\tilde{x}_{23}$,

$(11 + 0\theta, \delta_{324}(1 - \alpha), \delta_{324}(1 - \alpha))$ represents the value of the 4th basic variable i.e., $\tilde{x}_{32}$ and.

$(14 + 0\theta, \delta_{344}(1 - \alpha), \delta_{344}(1 - \alpha))$ represents the value of the 4th basic variable i.e., $\tilde{x}_{34}$ in a more-for-less solution of the symmetric triangular fuzzy balanced transportation problem represented by Table 2.5.

For example,

(a) Substituting $\theta = 9$, $\delta_{ij4} = 1 \forall i, j$, the obtained values of basic variables in a more-for-less solution are

$$\tilde{x}_{11} =_{Mag} (20, 1(1 - \alpha), 1(1 - \alpha)),$$

$$\tilde{x}_{13} =_{Mag} (0, 1(1 - \alpha), 1(1 - \alpha)),$$

$$\tilde{x}_{22} =_{Mag} (2, 1(1 - \alpha), 1(1 - \alpha)),$$

$$\tilde{x}_{23} =_{Mag} (17, 1(1 - \alpha), 1(1 - \alpha)),$$

$$\tilde{x}_{32} =_{Mag} (11, 1(1 - \alpha), 1(1 - \alpha)),$$

$\tilde{x}_{34} =_{Mag} (14, 1(1 - \alpha), 1(1 - \alpha))$ and the corresponding symmetric triangular fuzzy transportation cost is $\tilde{Z} =_{Mag} (143, 2(1 - \alpha), 2(1 - \alpha))$.

(b) Substituting $\theta = 9$, $\delta_{ij4} = 5 \forall i, j$, the obtained values of basic variables in a more-for-less solution are

$$\tilde{x}_{11} =_{Mag} (20, 5(1 - \alpha), 5(1 - \alpha)),$$

$$\tilde{x}_{13}=_{Mag}(0, 5(1-\alpha), 5(1-\alpha)),$$

$$\tilde{x}_{22}=_{Mag}(2, 5(1-\alpha), 5(1-\alpha)),$$

$$\tilde{x}_{23}=_{Mag}(17, 5(1-\alpha), 5(1-\alpha)),$$

$$\tilde{x}_{32}=_{Mag}(11, 5(1-\alpha), 5(1-\alpha)),$$

$\tilde{x}_{34}=_{Mag}(14, 5(1-\alpha), 5(1-\alpha))$ and the corresponding symmetric triangular fuzzy transportation cost is $\tilde{Z}=_{Mag}(143, 5(1-\alpha), 5(1-\alpha))$.

2.10 Results and Discussion

The results of the existing crisp balanced transportation problem [1, 2, 4], represented by Table 2.4 and the symmetric triangular fuzzy balanced transportation problem, represented by Table 2.5, obtained by the existing methods [1, 2, 4] and the proposed Mehar method-I, are shown in Table 2.7.

It is obvious from Table 2.7 that.

(i) On solving the existing problem, represented by Table 2.4, same more-for-less solutions are obtained by the existing methods [1, 2, 4] as well as by the proposed Mehar method-I.

(ii) The existing methods [1, 2, 4] cannot be used to solve the considered problem, represented by Table 2.5, as the existing methods [1, 2, 4] are proposed by considering the assumption that each parameter is a positive real number. While, each parameter of the considered problem, represented by Table 2.5, is a symmetric triangular fuzzy number. However, for the same problem, infinite number of more-for-less solutions (on substituting different values of $\theta \in (0, 9]$ in the mentioned more-for-less solution) are obtained by the proposed Mehar method-I.

2.11 Conclusions

In this chapter, the sufficient condition-I is proposed to check the existence of at least one more-for-less solution of symmetric triangular fuzzy balanced transportation problems. Also, a method is proposed to find all more-for-less solutions (if exist) of symmetric triangular fuzzy balanced transportation problems.

Table 2.7 Results of existing problem and considered problem

	More-for-less solutions	
	Existing problem [1, 2, 4]	Considered problem
Existing methods [1, 2, 4]	(i) For any value θ such that $0 < \theta \leq 5$, $x_{11}^2 = 0, x_{12}^2 = 0, x_{13}^2 = 7 + \theta, x_{14}^2 = 0, x_{15}^2 = 0, x_{21}^2 = 4 + 0\theta, x_{22}^2 = 6 + 0\theta, x_{23}^2 = 0, x_{24}^2 = 8 + 0\theta, x_{25}^2 = 0, x_{31}^2 = 0, x_{32}^2 = 0, x_{33}^2 = 5 - \theta, x_{34}^2 = 0, x_{35}^2 = 1 + \theta, x_{41}^2 = 0, x_{42}^2 = 5 + \theta, x_{43}^2 = 0, x_{44}^2 = 0, x_{45}^2 = 10 - \theta$ (ii) For any value of θ such that $0 < \theta \leq 5, x_{11}^2 = 0, x_{12}^2 = 0, x_{13}^2 = 7 + \theta, x_{14}^2 = 0, x_{15}^2 = 0, x_{21}^2 = 4 + 0\theta, x_{22}^2 = 6 - \theta, x_{23}^2 = 0, x_{24}^2 = 8 + \theta, x_{25}^2 = 0, x_{31}^2 = 0, x_{32}^2 = 0, x_{33}^2 = 5 - \theta, x_{34}^2 = 0, x_{35}^2 = 1 + \theta, x_{41}^2 = 0, x_{42}^2 = 5 + \theta, x_{43}^2 = 0, x_{44}^2 = 0, x_{45}^2 = 10 - \theta$ (iii) For any value θ such that $0 < \theta \leq 10$, $x_{11}^2 = 0, x_{12}^2 = 0, x_{13}^2 = 7 + 0\theta, x_{14}^2 = 0, x_{15}^2 = 0, x_{21}^2 = 4 + 0\theta, x_{22}^{2s} = 6 + 0\theta, x_{23}^2 = 0, x_{24}^2 = 8 + 0\theta, x_{25}^2 = 0, x_{31}^2 = 0, x_{32}^2 = 0, x_{33}^2 = 5 + 0\theta, x_{34}^2 = 0, x_{35}^2 = 1 + \theta, x_{41}^2 = 0, x_{42}^2 = 5 + \theta, x_{43}^2 = 0, x_{44}^2 = 0, x_{45}^2 = 10 - \theta$	Not applicable

(continued)

Table 2.7 (continued)

	More-for-less solutions	
	Existing problem [1, 2, 4]	Considered problem
Proposed Mehar method-I	(i) For any value θ such that $0 < \theta \leq 5$, $x_{11}^2 = 0, x_{12}^2 = 0, x_{13}^2 = 7 + \theta, x_{14}^2 = 0, x_{15}^2 = 0, x_{21}^2 = 4 + 0\theta, x_{22}^2 = 6 + 0\theta, x_{23}^2 = 0, x_{24}^2 = 8 + 0\theta, x_{25}^2 = 0, x_{31}^2 = 0, x_{32}^2 = 0, x_{33}^2 = 5 - \theta, x_{34}^2 = 0, x_{35}^2 = 1 + \theta, x_{41}^2 = 0, x_{42}^2 = 5 + \theta, x_{43}^2 = 0, x_{44}^2 = 0, x_{45}^2 = 10 - \theta$ (ii) For any value of θ such that $0 < \theta \leq 5, x_{11}^2 = 0, x_{12}^2 = 0, x_{13}^2 = 7 + \theta, x_{14}^2 = 0, x_{15}^2 = 0, x_{21}^2 = 4 + 0\theta, x_{22}^2 = 6 - \theta, x_{23}^2 = 0, x_{24}^2 = 8 + \theta, x_{25}^2 = 0, x_{31}^2 = 0, x_{32}^2 = 0, x_{33}^2 = 5 - \theta, x_{34}^2 = 0, x_{35}^2 = 1 + \theta, x_{41}^2 = 0, x_{42}^2 = 5 + \theta, x_{43}^2 = 0, x_{44}^2 = 0, x_{45}^2 = 10 - \theta$ (iii) For any value θ such that $0 < \theta \leq 10$, $x_{11}^2 = 0, x_{12}^2 = 0, x_{13}^2 = 7 + 0\theta, x_{14}^2 = 0, x_{15}^2 = 0, x_{21}^2 = 4 + 0\theta, x_{22}^2 = 6 + 0\theta, x_{23}^2 = 0, x_{24}^2 = 8 + 0\theta, x_{25}^2 = 0, x_{31}^2 = 0, x_{32}^2 = 0, x_{33}^2 = 5 + 0\theta, x_{34}^2 = 0, x_{35}^2 = 1 + \theta, x_{41}^2 = 0, x_{42}^2 = 5 + \theta, x_{43}^2 = 0, x_{44}^2 = 0, x_{45}^2 = 10 - \theta$	For any value θ such that $0 < \theta \leq 9$, $\tilde{x}_{11} = (11 + \theta, \delta_{114}(1 - \alpha), \delta_{114}(1 - \alpha)), \tilde{x}_{12} = (0, \delta_{124}(1 - \alpha), \delta_{124}(1 - \alpha)), \tilde{x}_{13} = (9 - \theta, \delta_{134}(1 - \alpha), \delta_{134}(1 - \alpha)), \tilde{x}_{14} = (0, \delta_{144}(1 - \alpha), \delta_{144}(1 - \alpha)), \tilde{x}_{21} = (0, \delta_{214}(1 - \alpha), \delta_{214}(1 - \alpha)), \tilde{x}_{22} = (2 + 0\theta, \delta_{224}(1 - \alpha), \delta_{224}(1 - \alpha)), \tilde{x}_{23} = (8 + \theta, \delta_{234}(1 - \alpha), \delta_{234}(1 - \alpha)), \tilde{x}_{24} = (0, \delta_{244}(1 - \alpha), \delta_{244}(1 - \alpha)), \tilde{x}_{31} = (0, \delta_{314}(1 - \alpha), \delta_{314}(1 - \alpha)), \tilde{x}_{32} = (11 + 0\theta, \delta_{324}(1 - \alpha), \delta_{324}(1 - \alpha)), \tilde{x}_{33} = (0, \delta_{334}(1 - \alpha), \delta_{334}(1 - \alpha)), \tilde{x}_{34} = (14 + 0\theta, \delta_{344}(1 - \alpha), \delta_{344}(1 - \alpha))$

References

1. V. Adlakha, K. Kowalski, A note on the procedure MFL for a more-for-less solution in transportation problems. Omega **28**, 481–483 (2000)
2. H. Arsham, Postoptimality analyses of the transportation problem. J. Operat. Res. Soc. **43**, 121–139 (1992)
3. C.R. Bector, S. Chandra, *Fuzzy mathematical programming and fuzzy matrix games* (Springer, Berlin, Heidelberg, New York, 2005)
4. W. Szwarc, The transportation paradox. Naval Res. Log. Quart. **18**, 185–202 (1971)
5. H.A. Taha, Operations Research: An Introduction. Pearson, Prentice Hall, Upper Saddle River, NJ (2013)
6. V. Vidhya, P. Uma Maheswari, K. Ganesan, An alternate method for finding more for less solution to fuzzy transportation problem with mixed constraints. Soft Comp. **25**, 11989–11996 (2021)

Chapter 3
Mehar Method-II to Find All More-For-Less Solutions of Symmetric Fuzzy Transportation Problems with Mixed Constraints

In this chapter, a new method (named as Mehar method-II) is proposed to check the existence of at least one more-for-less solution as well as to find all more-for-less solutions (if exists) of symmetric triangular fuzzy transportation problems with mixed constraints (transportation problems with mixed constraints in which each parameter and each variable is represented by a symmetric triangular fuzzy number). Also, to illustrate the proposed Mehar method-II, an existing crisp transportation problem with mixed constraints and an existing symmetric triangular fuzzy transportation problem with mixed constraints are solved by the proposed Mehar method-II. Furthermore, the superiority of the proposed Mehar method-II over some existing methods is discussed.

3.1 Tabular Representation of Crisp Transportation Problems with Mixed Constraints

In the literature [1, 3, 4], it is pointed out that in general, to find an optimal solution of a transportation problem, it is assumed that (i) the total quantity of the product, transported from a source at different destinations, will be at most equal to the availability of the product at that source (ii) the total quantity of the product, received at a destination from different sources, will be at least equal to the demand of the product at that destination. However, if it is assumed that (i) the total quantity of the product, transported from a source at different destinations, may also be more than the availability of the product at that source (ii) the total quantity of the product, received at a destination from different sources, may also be less than the demand of the product at that destination. Then, such a transportation problem is known as transportation problem with mixed constraints.

A crisp transportation problem with mixed constraints can be represented into tabular form as shown by Table 3.1.

T. K. Bhatia et al., *More-for-Less Solutions in Fuzzy Transportation Problems*, Studies in Fuzziness and Soft Computing 426,
https://doi.org/10.1007/978-3-031-30337-1_3

Table 3.1 Tabular representation of a crisp transportation problem with mixed constraints

	D_1	D_2	...	D_n	Availability
S_1	c_{11}	c_{12}	...	c_{1n}	$= or \leq or \geq a_1$
S_2	c_{21}	c_{22}	...	c_{2n}	$= or \leq or \geq a_2$
$\vdots$	$\vdots$	$\vdots$	$\vdots$	$\vdots$	$\vdots$
S_m	c_{m1}	c_{m2}	...	c_{mn}	$= or \leq or \geq a_m$
Demand	$= or \leq or \geq b_1$	$= or \leq or \geq b_2$	...	$= or \leq or \geq b_n$	

where,

(i) m represents the number of sources.
(ii) n represents the number of destinations.
(iii) S_i represents the ith source.
(iv) D_j represents the jth destination.
(v) The positive real number c_{ij} represents the cost for transporting one unit quantity of the product from the ith source to the jth destination.
(vi) The positive real number a_i represents the availability of the product at the ith source.
(vii) The positive real number b_j represents the demand of the product at the jth destination.

3.2 Tabular Representation of Symmetric Triangular Fuzzy Transportation Problems with Mixed Constraints

A symmetric triangular fuzzy transportation problem with mixed constraints can be represented into tabular form as shown by Table 3.2.

Table 3.2 Tabular representation of a symmetric fuzzy transportation problem with mixed constraints

	D_1	D_2	...	D_n	Availability
S_1	$\tilde{c}_{11}$	$\tilde{c}_{12}$	...	$\tilde{c}_{1n}$	$=_{Mag} or \leq_{Mag} or \geq_{Mag} \tilde{a}_1$
S_2	$\tilde{c}_{21}$	$\tilde{c}_{22}$	...	$\tilde{c}_{2n}$	$=_{Mag} or \leq_{Mag} or \geq_{Mag} \tilde{a}_2$
$\vdots$	$\vdots$	$\vdots$	$\vdots$	$\vdots$	$\vdots$
S_m	$\tilde{c}_{m1}$	$\tilde{c}_{m2}$	...	$\tilde{c}_{mn}$	$=_{Mag} or \leq_{Mag} or \geq_{Mag} \tilde{a}_m$
Demand	$=_{Mag} or \leq_{Mag} or \geq_{Mag} \tilde{b}_1$	$=_{Mag} or \leq_{Mag} or \geq_{Mag} \tilde{b}_2$	...	$=_{Mag} or \leq_{Mag} or \geq_{Mag} \tilde{b}_n$	

where,

(i) The symmetric triangular fuzzy number $\tilde{c}_{ij} = \left(c^2_{ij}, \delta_{ij1}(1-\alpha), \delta_{ij1}(1-\alpha)\right)$ represents the fuzzy cost for transporting one unit quantity of product from the ith source to the jth destination.
(ii) The symmetric triangular fuzzy number $\tilde{a}_i = \left(a^2_i, \delta_{i2}(1-\alpha), \delta_{i2}(1-\alpha)\right)$ represents the fuzzy supply of product at the ith source.
(iii) The symmetric triangular fuzzy number $\tilde{b}_j = \left(b^2_j, \delta_{j3}(1-\alpha), \delta_{j3}(1-\alpha)\right)$ represents the fuzzy demand of product at the jth destination.

3.3 Crisp Linear Programming Problems Corresponding to Crisp Transportation Problems with Mixed Constraints

If it is assumed that the non-negative real number x_{ij} represents the quantity of the product to be supplied from the ith source to the jth destination and Z represents the total crisp transportation cost. Then, the crisp transportation problem with mixed constraints, represented by Table 3.1, can be formulated into its equivalent crisp linear programming problem (P3.1) or into its equivalent crisp linear programming problem (P3.2) (dual problem of the crisp linear programming problem (P3.1).

Problem (P3.1) *Minimize* $\left(Z = \sum_{i=1}^{m}\sum_{j=1}^{n} c_{ij}x_{ij}\right)$
Subject to

$$\sum_{j=1}^{n} x_{ij} = or \geq or \leq a_i;\ i = 1, 2, \ldots, m;$$

(ith source constraint)

$$\sum_{i=1}^{m} x_{ij} = or \geq or \leq b_j;\ j = 1, 2, \ldots, n;$$

(jth destination constraint)

$$x_{ij} \geq 0;\ i = 1, 2, \ldots, m;\ j = 1, 2, \ldots, n.$$

Problem (P3.2) *Maximize* $\left(Z = \sum_{i=1}^{m} a_i u_i + \sum_{j=1}^{n} b_j v_j\right)$
Subject to

$$u_i + v_j \le c_{ij};\ i = 1, 2, \ldots, m;\ j = 1, 2, \ldots, n;$$

$$u_i * 0,\ v_j \circ 0;\ i = 1, 2, \ldots, m;\ j = 1, 2, \ldots, n;$$

where,

(i) $*$ will be $\ge$ if the sign of the ith source constraint of the crisp linear programming problem ($P3.1$) is $\ge$.
(ii) $*$ will be $\le$ if the sign of the ith source constraint of the crisp linear programming problem ($P3.1$) is $\le$.
(iii) If the sign of the ith source constraint of the crisp linear programming problem ($P3.1$) is $=$. Then, u_i will be an unrestricted variable.
(iv) $\circ$ will be $\ge$ if the sign of the jth destination constraint of the crisp linear programming problem ($P3.1$) is $\ge$.
(v) $\circ$ will be $\le$ if the sign of the jth destination constraint of the crisp linear programming problem ($P3.1$) is $\le$.
(vi) If the sign of the $j$thdestination constraint of the crisp linear programming problem ($P3.1$) $=$. Then, v_j will be an unrestricted variable.

It is pertinent to mention that the number of linearly independent constraints in the crisp linear programming problem ($P3.1$) may be $(m + n)$ (Please see Sect. 3.9.3.2 for more details). Hence, an optimal solution of the crisp transportation problem with mixed constraints, represented by Table 3.1, cannot be obtained by deleting any one of the constraints of the crisp linear programming problem ($P3.1$).

3.4 Fuzzy Linear Programming Problems Corresponding to Symmetric Triangular Fuzzy Transportation Problems with Mixed Constraints

If it is assumed that the symmetric triangular fuzzy number $\tilde{x}_{ij} = \left(x_{ij}^2, \delta_{ij4}(1-\alpha), \delta_{ij4}(1-\alpha)\right)$ represents the fuzzy quantity of the product to be supplied from the ith source to the jth destination, the symmetric triangular fuzzy number $\tilde{Z}$ represents the total fuzzy transportation cost and $\tilde{0}$ represents the zero triangular fuzzy number. Then, the symmetric triangular fuzzy transportation problem with mixed constraints, represented by Table 3.2, can be formulated into its equivalent fuzzy linear programming problem ($P3.3$) or into its equivalent crisp linear programming problem ($P3.4$) (dual problem of the fuzzy linear programming problem ($P3.3$)).

Problem ($P3.3$) *Minimize* $\left(\tilde{Z} =_{Mag} \oplus_{i=1}^{m} \oplus_{j=1}^{n} \left(\tilde{c}_{ij} \otimes \tilde{x}_{ij}\right)\right)$
Subject to

$$\oplus_{j=1}^{n}\tilde{x}_{ij} =_{Mag} or \leq_{Mag} or \geq_{Mag} \tilde{a}_i;\ i = 1, 2, \ldots, m$$

; (*i*th source constraint)

$$\oplus_{i=1}^{m}\tilde{x}_{ij} =_{Mag} or \leq_{Mag} or \geq_{Mag} \tilde{b}_j;\ j = 1, 2, \ldots, n;$$

(*j*th destination constraint)

$$\tilde{x}_{ij} \geq_{Mag} \tilde{0};\ i = 1, 2, \ldots, m;\ j = 1, 2, \ldots, n.$$

Problem (*P*3.4) *Maximize* $\left(Z^2 = \sum_{i=1}^{m} a_i^2 u_i^2 + \sum_{j=1}^{n} b_j^2 v_j^2\right)$

Subject to

$$u_i^2 + v_j^2 \leq c_{ij}^2;\ i = 1, 2, \ldots, m;\ j = 1, 2, \ldots, n;$$

$$u_i^2 * 0,\ v_j^2 \circ 0;\ i = 1, 2, \ldots, m;\ j = 1, 2, \ldots, n;$$

where,

(i) $*$ will be $\geq$ if the sign of the *i*th source constraint of the fuzzy linear programming problem (*P*3.3) is $\geq$.
(ii) $*$will be $\leq$ if the sign of the *i*th source constraint of the fuzzy linear programming problem (*P*3.3) is $\leq$.
(iii) If the sign of the *i*th source constraint of the fuzzy linear programming problem (*P*3.3) is $=$. Then, u_i will be an unrestricted variable.
(iv) $\circ$ will be $\geq$ if the sign of the *j*th destination constraint of the fuzzy linear programming problem (*P*3.3) is $\geq$.
(v) $\circ$ will be $\leq$ if the sign of the *j*th destination constraint of the fuzzy linear programming problem (*P*3.3) is $\leq$.
(vi) If the sign of the *j*th destination constraint of the fuzzy linear programming problem (*P*3.3) $=$. Then, v_j will be an unrestricted variable.

It is pertinent to mention that the number of linearly independent constraints in the fuzzy linear programming problem (*P*3.3) may be $(m + n)$ (Please see Sect. 3.9.3.1 for more details). Hence, an optimal solution of the symmetric triangular fuzzy transportation problem with mixed constraints, represented by Table 3.2, cannot be obtained by deleting any one of the constraints of the fuzzy linear programming problem (*P*3.3).

The dual problem (*P*3.4) is obtained as follows.

Step 1: Using Definition 2.1.10, discussed in Chap. 2, the fuzzy linear programming problem (*P*3.3) can be transformed into its equivalent crisp linear programming problem (*P*3.5).

Problem (*P*3.5) *Minimize* $\left(Mag\left(\tilde{Z}\right) = Mag\left(\oplus_{i=1}^{m} \oplus_{j=1}^{n} \left(\tilde{c}_{ij} \otimes \tilde{x}_{ij}\right)\right)\right)$

Subject to

$$Mag\left(\oplus_{j=1}^{n}\tilde{x}_{ij}\right) = or \geq or \leq Mag(\tilde{a}_i);\ i = 1, 2, \ldots, m;$$

$$Mag\left(\oplus_{i=1}^{m}\tilde{x}_{ij}\right) = or \geq or \leq Mag\left(\tilde{b}_j\right);\ j = 1, 2, \ldots, n;$$

$$Mag\left(\tilde{x}_{ij}\right) \geq Mag\left(\tilde{0}\right);\ i = 1, 2, \ldots, m;\ j = 1, 2, \ldots, n.$$

Step 2: Using the expression $Mag\left(\oplus_{i=1}^{k}\tilde{A}_i\right) = \sum_{i=1}^{k} Mag\left(\tilde{A}_i\right)$, discussed in Chap. 2, the crisp linear programming problem ($P3.5$) can be transformed into its equivalent crisp linear programming problem ($P3.6$).

Problem ($P3.6$) *Minimize* $\left(Mag\left(\tilde{Z}\right) = \left(\sum_{i=1}^{m}\sum_{j=1}^{n} Mag\left(\tilde{c}_{ij} \otimes \tilde{x}_{ij}\right)\right)\right)$

Subject to

$$\sum_{j=1}^{n} Mag\left(\tilde{x}_{ij}\right) = or \geq or \leq Mag(\tilde{a}_i);\ i = 1, 2, \ldots, m;$$

$$\sum_{i=1}^{m} Mag\left(\tilde{x}_{ij}\right) = or \geq or \leq Mag\left(\tilde{b}_j\right);\ j = 1, 2, \ldots, n;$$

$$Mag\left(\tilde{x}_{ij}\right) \geq Mag\left(\tilde{0}\right);\ i = 1, 2, \ldots, m;\ j = 1, 2, \ldots, n.$$

Step 3: Using the expression $Mag\left(\otimes_{i=1}^{k}\tilde{A}_i\right) = \prod_{i=1}^{k} Mag\left(\tilde{A}_i\right)$, discussed in Chap. 2, the crisp linear programming problem ($P3.6$) can be transformed into its equivalent crisp linear programming problem ($P3.7$).

Problem ($P3.7$) *Minimize* $\left(Mag\left(\tilde{Z}\right) = \left(\sum_{i=1}^{m}\sum_{j=1}^{n} Mag\left(\tilde{c}_{ij}\right)Mag\left(\tilde{x}_{ij}\right)\right)\right)$

Subject to

Constraints of the problem ($P3.6$).

Step 4: Using the expression $Mag\left(\tilde{A}_i\right) = a_i^2$, discussed in Chap. 2, the crisp linear programming problem ($P3.7$) can be transformed into its equivalent crisp linear programming problem ($P3.8$).

Problem ($P3.8$) *Minimize* $\left(Z^2 = \sum_{i=1}^{m}\sum_{j=1}^{n} c_{ij}^2 x_{ij}^2\right)$

Subject to

$$\sum_{j=1}^{n} x_{ij}^2 = or \geq or \leq a_i^2;\ i = 1, 2, \ldots, m;$$

$$\sum_{i=1}^{m} x_{ij}^2 = or \geq or \leq b_j^2;\ j = 1, 2, \ldots, n;$$

$$x_{ij}^2 \geq 0;\ i = 1, 2, \ldots, m;\ j = 1, 2, \ldots, n.$$

Step 5: Using the duality theory [5], the crisp linear programming problem ($P3.4$) represents the dual of the crisp linear programming problem ($P3.8$). Hence, the crisp linear programming problem ($P3.4$) represents the dual of the fuzzy linear programming problem ($P3.3$).

3.5 Crisp Transportation Problems with Mixed Constraints Equivalent to Symmetric Triangular Fuzzy Transportation Problems with Mixed Constraints

The following clearly indicates that the symmetric triangular fuzzy transportation problem with mixed constraints, represented by Table 3.2, is equivalent to crisp transportation problem with mixed constraints represented by Table 3.3.

It is obvious from Step 1 to Step 4 of Sect. 3.4 that the fuzzy linear programming problem ($P3.3$) corresponding to the symmetric triangular fuzzy transportation problem with mixed constraints, represented by Table 3.2, is equivalent to the crisp linear programming problem ($P3.8$). Also, it is obvious from Sect. 3.3 that the crisp linear programming problem ($P3.8$) is equivalent to crisp transportation problem with mixed constraints represented by Table 3.3.

Table 3.3 Tabular representation of a crisp transportation problem with mixed constraints

	D_1	D_2	...	D_n	Availability
S_1	c_{11}^2	c_{12}^2	...	c_{1n}^2	$= or \leq or \geq a_1^2$
S_2	c_{21}^2	c_{22}^2	...	c_{2n}^2	$= or \leq or \geq a_2^2$
$\vdots$	$\vdots$	$\vdots$	$\vdots$	$\vdots$	$\vdots$
S_m	c_{m1}^2	c_{m2}^2	...	c_{mn}^2	$= or \leq or \geq a_m^2$
Demand	$= or \leq or \geq b_1^2$	$= or \leq or \geq b_2^2$	...	$= or \leq or \geq b_n^2$	

3.6 Proposed Sufficient Condition-II for the Existence of at Least One More-For-Less Solution

It is obvious from Sect. 3.5 that the symmetric triangular fuzzy transportation problem with mixed constraints, represented by Table 3.2, is equivalent to the crisp transportation problem with mixed constraints represented by Table 3.3. Hence, to propose a sufficient condition-II for the existence of at least one more-less-solution for the symmetric triangular fuzzy transportation problem with mixed constraints, represented by Table 3.2, is equivalent to propose a sufficient condition-II for the existence of at least one more-less-solution for the crisp transportation problem with mixed constraints represented by Table 3.3. Keeping the same in mind, in this section, the following sufficient condition-II for the existence of at least one more-for-less solution of crisp transportation problem with mixed constraints, represented by Table 3.3, is proposed.

> "If in an optimal solution of the crisp linear programming problem ($P3.4$), there exist variables u_p^2 and v_q^2 such that $u_p^2 + v_q^2 < 0$. Then, at least one more-for-less solution of crisp transportation problem with mixed constraints, represented by Table 3.3, will exist. Hence, at least one more-for-less solution of symmetric triangular fuzzy transportation problem with mixed constraints, represented by Table 3.2, will exist."

This sufficient condition-II is obtained as follows.

According to the duality theory, to find the total minimum transportation cost for the crisp transportation problem with mixed constraints, represented by Table 3.3, is equivalent to find the optimal value of the objective function i.e., $Z^2 = \sum_{i=1}^{m}\sum_{j=1}^{n} c_{ij}^2 x_{ij}^2$ of the crisp linear programming problem ($P3.8$) or the optimal value of the objective function i.e., $Z^2 = \sum_{i=1}^{m} a_i^2 u_i^2 + \sum_{j=1}^{n} b_j^2 v_j^2$ of the dual problem ($P3.4$) of the crisp linear programming problem ($P3.8$).

Increasing the availability of the pth source and the demand of the qth destination by the positive quantity θ, the total crisp transportation cost of the modified problem is

$$Z^2_{modified} = \sum_{\substack{i=1 \\ i \neq p}}^{m} a_i^2 u_i^2 + (a_p^2 + \theta) u_p^2 + \sum_{\substack{j=1 \\ j \neq q}}^{n} b_j^2 v_j^2 + (b_q^2 + \theta) v_q^2$$

$$Z^2_{modified} = \sum_{\substack{i=1 \\ i \neq p}}^{m} a_i^2 u_i^2 + (a_p^2 u_p^2 + \theta u_p^2) + \sum_{\substack{j=1 \\ j \neq q}}^{n} b_j^2 v_j^2 + (b_q^2 v_q^2 + \theta v_q^2)$$

$$Z^2_{modified} = \sum_{i=1}^{m} a_i^2 u_i^2 + \theta u_p^2 + \sum_{j=1}^{n} b_j^2 v_j^2 + \theta v_q^2 \tag{3.1}$$

Substituting $Z^2 = \sum_{i=1}^{m} a_i^2 u_i^2 + \sum_{j=1}^{n} b_j^2 v_j^2$ in (3.1),

$$Z^2_{modified} = Z^2 + \theta u_p^2 + \theta v_q^2$$

Since, θ is a positive real number. So, the total crisp transportation cost i.e., $Z^2_{modified} = Z^2 + \theta u_p^2 + \theta v_q^2$ of the modified problem will be less than the total crisp transportation cost i.e., Z^2 only if $u_p^2 + v_q^2 < 0$.

3.7 Proposed Mehar Method-II

In this section, a new method (named as Mehar method-II) is proposed to check the existence of at least one more-for-less solution as well as to find all more-for-less solutions (if exists) of symmetric triangular fuzzy transportation problems with mixed constraints.

The steps of the proposed Mehar method-II are as follows.

Step 1: Use **Step 1.1** to **Step 1.4** to check that for the symmetric triangular fuzzy transportation problem with mixed constraints, represented by Table 3.2, at least one more-for-less solution will exist or not.

Step 1.1: Transform the symmetric triangular fuzzy transportation problem with mixed constraints, represented by Table 3.2, into its equivalent crisp transportation problem with mixed constraints represented by Table 3.3.
Step 1.2: Find an optimal solution $\left\{u_i^2; i = 1, 2, \ldots, m, v_j^2; j = 1, 2, \ldots, n\right\}$ of the problem (P3.4) by substituting the values of c_{ij}^2, a_i^2 and b_j^2 from Table 3.3.
Step 1.3: Using the optimal values of $u_i^2; i = 1, 2, \ldots, m, v_j^2; j = 1, 2, \ldots, n$, obtained in **Step 1.2**, construct a shadow price matrix $\left[A_{ij}\right]_{m \times n}$, where $A_{ij} = u_i^2 + v_j^2$.
Step 1.4: Check that all the elements A_{ij} of the shadow price matrix, obtained in **Step 1.3**, are non-negative or not.

Case (i): If all the elements of the shadow price matrix, obtained in **Step 1.3**, are non-negative. Then, there does not exist a more-for-less solution.
Case (ii): If at least one element of the shadow price matrix, obtained in **Step 1.3**, is negative. Then, at least one more-for-less solution will exist.

Step 2: If the element corresponding to the pth row and the qth column of the shadow price matrix, obtained in **Step 1.3**, is negative i.e., if $u_p^2 + v_q^2 < 0$. Then, a more-for-less solution can be obtained by increasing the availability of the pth source and the demand of the qth destination with the help of **Step 2.1** to **Step 2.8** of the post-optimality analysis [5].

Step 2.1: Construct the matrix

$$R_{pq} = \begin{bmatrix} a_1 \\ a_2 \\ \vdots \\ a_p + \theta \\ \vdots \\ a_m \\ b_1 \\ b_2 \\ \vdots \\ b_q + \theta \\ \vdots \\ b_n \end{bmatrix}$$

where,

(i) θ is a positive real number.
(ii) a_i is the availability of the product at the ith source.
(iii) b_i is the demand of the product at the jth destination.

Step 2.2: Identify the basic variables in an optimal solution of the crisp linear programming problem ($P3.9$).

Problem (*P*3.9) *Minimize* $\left(Z^2 = \sum_{i=1}^{m} \sum_{j=1}^{n} c_{ij}^2 x_{ij}^2\right)$
Subject to

$$\sum_{j=1}^{n} x_{ij}^2 + P_i s_i^2 = a_i^2;\ i = 1, 2, \ldots, m;$$

$$\sum_{i=1}^{m} x_{ij}^2 + Q_j s_j^2 = b_j^2;\ j = 1, 2, \ldots, n;$$

$$x_{ij}^2, s_i^2, s_j^2 \geq 0;\ i = 1, 2, \ldots, m;\ j = 1, 2, \ldots, n,$$

where,

(i) $P_i = 1$ if the sign of the ith source constraint of the crisp linear programming problem ($P3.8$) is $\leq$.

(ii) $P_i = -1$ if the sign of the ith source constraint of the crisp linear programming problem ($P3.8$) is $\geq$.

(iii) $P_i = 0$ if the sign of the ith source constraint of the crisp linear programming problem ($P3.8$) is $=$.

(iv) $Q_j = 1$ if the sign of the jth destination constraint of the crisp linear programming problem ($P3.8$) is $\leq$.

(v) $Q_j = -1$ if the sign of the jth destination constraint of the crisp linear programming problem ($P3.8$) is $\geq$.

(vi) $Q_j = 0$ if the sign of the jth destination constraint of the crisp linear programming problem ($P3.8$) is $=$.

Step 2.3: Construct a basis matrix B with the help of the coefficients of the identified optimal basic variables in the constraints of the crisp linear programming problem ($P3.9$).

Step 2.4: Find the multiplicative inverse of the basis matrix B i.e., find B^{-1}.

Step 2.5: Find the matrix $B^{-1}R_{pq}$.

Step 2.6: Find the range of θ(where, $\theta > 0$) for which all the elements of the matrix $B^{-1}R_{pq}$, obtained in **Step 2.5**, are greater than or equal to zero.

Step 2.7: For any value of θ lying in its range, obtained in **Step 2.6**, the ith element (say, τ_i) of the matrix $B^{-1}R_{pq}$ represents the value of the ith basic variable in a more-for-less solution of crisp transportation problem with mixed constraints represented by Table 3.3.

Step 2.8: For any value of θ lying in its range, obtained in **Step 2.6**, the symmetric triangular fuzzy number $(\tau_i, \delta_i(1-\alpha), \delta_i(1-\alpha))$, where δ_i is any non-negative real number and $\alpha \in (0, 1]$, represents the value of the ith basic fuzzy variable in a more-for-less solution of the symmetric triangular fuzzy transportation problem with mixed constraints represented by Table 3.2.

3.8 All More-For-Less Solutions of Existing Problems

In this section, all more-for-less solutions of the existing crisp transportation problem with mixed constraints [2] as well as the existing symmetric triangular fuzzy transportation problem with mixed constraints [6] are obtained by the proposed Mehar method-II.

3.8.1 All More-For-Less Solutions of the First Problem

Using the proposed Mehar method-II, all more-for-less solutions (if exists) of the existing crisp transportation problem with mixed constraints, represented by Table 3.4, can be obtained as follows.

Table 3.4 Crisp transportation problem with mixed constraints [2]

	D_1	D_2	D_3	Availability
S_1	10	1	4	$= 5$
S_2	5	7	1	≥ 6
S_3	8	9	2	≤ 9
Demand	$= 8$	≥ 10	≤ 5	

Step 1: According to **Step 1** of the proposed Mehar method-II, there is a need to use **Step 1.1** to **Step 1.4** to check that for the crisp transportation problem with mixed constraints, represented by Table 3.4, at least one more-for-less solution will exist or not.

Step 1.1: According to **Step 1.1** of the proposed Mehar method-II, there is a need to transform the symmetric triangular fuzzy transportation problem with mixed constraints into its equivalent crisp transportation problem with mixed constraints. Since, the transportation problem with mixed constraints, represented by Table 3.4, is a crisp transportation problem with mixed constraints. So, there is no need to apply this step.

Step 1.2: According to **Step 1.2** of the proposed Mehar method-II, there is a need to find an optimal solution of the crisp linear programming problem ($P3.10$).

Problem ($P3.10$)

$Maximize(5u_1^2 + 6u_2^2 + 9u_3^2 + 8v_1^2 + 10v_2^2 + 5v_3^2)$

Subject to

$$u_1^2 + v_1^2 \leq 10,$$

$$u_1^2 + v_2^2 \leq 1,$$

$$u_1^2 + v_3^2 \leq 4,$$

$$u_2^2 + v_1^2 \leq 5,$$

$$u_2^2 + v_2^2 \leq 7,$$

$$u_2^2 + v_3^2 \leq 1,$$

$$u_3^2 + v_1^2 \leq 8,$$

$$u_3^2 + v_2^2 \leq 9,$$

$$u_3^2 + v_3^2 \leq 2,$$

u_1^2, v_1^2 are unrestricted in sign,
$u_2^2, v_2^2 \geq 0,$
$u_3^2, v_3^2 \leq 0.$

It can be easily verified that on solving the problem ($P3.10$), the obtained optimal solution is $u_1^2 = -6, u_2^2 = 0, u_3^2 = 0, v_1^2 = 5, v_2^2 = 7, v_3^2 = 0$.

Step 1.3: According to **Step 1.3** of the proposed Mehar method-II, the following shadow price matrix is obtained.

$$\begin{bmatrix} u_1^2+v_1^2=2 & u_1^2+v_2^2=-1 & u_1^2+v_3^2=-4 \\ u_2^2+v_1^2=6 & u_2^2+v_2^2=3 & u_2^2+v_3^2=0 \\ u_3^2+v_1^2=6 & u_3^2+v_2^2=3 & u_3^2+v_3^2=0 \end{bmatrix}$$

Step 1.4: Since, some elements of the shadow price matrix, obtained in **Step 1.3**, are negative. So, according to Case (ii) of **Step 1.4** of the proposed Mehar method-II, there will exist at least one more-for-less solution of the crisp transportation problem with mixed constraints represented by Table 3.4.

Step 2: According to Step 2 of the proposed Mehar method-II, all more-for-less solutions of the crisp transportation problem with mixed constraints, represented by Table 3.4, can be obtained as follows.

Step 2.1: It is obvious from **Step 1.3** that $u_1^2+v_1^2<0$ and $u_1^2+v_3^2<0$. Therefore, according to **Step 2.1** of the proposed Mehar method-II,

(i) $R_{11}=\begin{bmatrix} 5+\theta \\ 6 \\ 9 \\ 8+\theta \\ 10 \\ 5 \end{bmatrix}$

(ii) $R_{13}=\begin{bmatrix} 5+\theta \\ 6 \\ 9 \\ 8 \\ 10 \\ 5+\theta \end{bmatrix}$

Step 2.2: According to **Step 2.2** of the proposed Mehar method-II, there is a need to identify the basic variables in an optimal solution of the crisp linear programming problem ($P3.11$).

Problem ($P3.11$)

$Minimize\left(Z=10x_{11}^2+x_{12}^2+4x_{13}^2+5x_{21}^2+7x_{22}^2+x_{23}^2+8x_{31}^2+9x_{32}^2+2x_{33}^2\right)$

Subject to

$$x_{11}^2+x_{12}^2+x_{13}^2=5,$$

$$x_{21}^2+x_{22}^2+x_{23}^2-s_1^2=6,$$

$$x_{31}^2+x_{32}^2+x_{33}^2+s_2^2=9,$$

$$x_{11}^2 + x_{21}^2 + x_{31}^2 = 8,$$

$$x_{12}^2 + x_{22}^2 + x_{32}^2 - s_3^2 = 10,$$

$$x_{13}^2 + x_{23}^2 + x_{33}^2 + s_4^2 = 5,$$

$$x_{ij}^2 \geq 0; i = 1, 2, 3; j = 1, 2, 3; s_k^2 \geq 0; k = 1, 2, 3, 4.$$

It can be easily verified that the following optimal solution is obtained on solving the crisp linear programming problem ($P3.11$).

$x_{11}^2 = 0, x_{12}^2 = 5, x_{13}^2 = 0, x_{21}^2 = 8, x_{22}^2 = 5, x_{23}^2 = 0, x_{31}^2 = 0, x_{32}^2 = 0, x_{33}^2 = 0, s_1^2 = 7, s_2^2 = 9, s_3^2 = 0, s_4^2 = 5.$

Also, it can be easily verified that the basic variables in this obtained optimal solution are $x_{12}^2, x_{21}^2, x_{22}^2, s_1^2, s_2^2$ and s_4^2.

Step 2.3: According to **Step 2.3** of the proposed Mehar method-II, there is a need to construct the basis matrix B with the help of the coefficients of optimal basic variables $x_{12}^2, x_{21}^2, x_{22}^2, s_1^2, s_2^2, s_4^2$ in the constraints of the crisp linear programming problem ($P3.11$).

$$B = \begin{bmatrix} 1 & 0 & 0 & 0 & 0 & 0 \\ 0 & 1 & 1 & -1 & 0 & 0 \\ 0 & 0 & 0 & 0 & 1 & 0 \\ 0 & 1 & 0 & 0 & 0 & 0 \\ 1 & 0 & 1 & 0 & 0 & 0 \\ 0 & 0 & 0 & 0 & 0 & 1 \end{bmatrix}$$

Step 2.4: According to **Step 2.4** of the proposed Mehar method-II, there is a need to find the multiplicative inverse of the basis matrix B.

$$B^{-1} = \begin{bmatrix} 1 & 0 & 0 & 0 & 0 & 0 \\ 0 & 0 & 0 & 1 & 0 & 0 \\ -1 & 0 & 0 & 0 & 1 & 0 \\ -1 & -1 & 0 & 1 & 1 & 0 \\ 0 & 0 & 1 & 0 & 0 & 0 \\ 0 & 0 & 0 & 0 & 0 & 1 \end{bmatrix}$$

Step 2.5: According to **Step 2.5** of the proposed Mehar method-II, there is a need to find $B^{-1}R_{11}$ and $B^{-1}R_{13}$.

(i) $$B^{-1}R_{11} = \begin{bmatrix} 1 & 0 & 0 & 0 & 0 & 0 \\ 0 & 0 & 0 & 1 & 0 & 0 \\ -1 & 0 & 0 & 0 & 1 & 0 \\ -1 & -1 & 0 & 1 & 1 & 0 \\ 0 & 0 & 1 & 0 & 0 & 0 \\ 0 & 0 & 0 & 0 & 0 & 1 \end{bmatrix} \begin{bmatrix} 5+\theta \\ 6 \\ 9 \\ 8+\theta \\ 10 \\ 5 \end{bmatrix} = \begin{bmatrix} 5+\theta \\ 8+\theta \\ 5-\theta \\ 7+0\theta \\ 9+0\theta \\ 5+0\theta \end{bmatrix}$$

(ii) $$B^{-1}R_{13} = \begin{bmatrix} 1 & 0 & 0 & 0 & 0 & 0 \\ 0 & 0 & 0 & 1 & 0 & 0 \\ -1 & 0 & 0 & 0 & 1 & 0 \\ -1 & -1 & 0 & 1 & 1 & 0 \\ 0 & 0 & 1 & 0 & 0 & 0 \\ 0 & 0 & 0 & 0 & 0 & 1 \end{bmatrix} \begin{bmatrix} 5+\theta \\ 6 \\ 9 \\ 8 \\ 10 \\ 5+\theta \end{bmatrix} = \begin{bmatrix} 5+\theta \\ 8+0\theta \\ 5-\theta \\ 7-\theta \\ 9+0\theta \\ 5+\theta \end{bmatrix}$$

Step 2.6: According to **Step 2.6** of the proposed Mehar method-II, there is a need to find the range of θ (where, $\theta > 0$) so that all the elements of the matrix, obtained in **Step 2.5**, are greater than or equal to zero.

It is obvious from **Step 2.5** that

(i) The range of θ for which all the elements of the matrix $B^{-1}R_{11}$, obtained in **Step 2.5**, are greater than or equal to zero, i.e., $5+\theta \geq 0, 8+\theta \geq 0, 5-\theta \geq 0$ is $0 < \theta \leq 5$.

(ii) The range of θ for which all the elements of the matrix $B^{-1}R_{13}$, obtained in **Step 2.5**, are greater than or equal to zero, i.e., $5+\theta \geq 0, 5-\theta \geq 0, 7-\theta \geq 0$ is $0 < \theta \leq 5$.

Step 2.7: According to **Step 2.7** of the proposed Mehar method-II,

(i) For any value θ such that $0 < \theta \leq 5$,

$5+\theta$ represents the value of the 1st basic variable i.e., x_{12}^2,
$8+\theta$ represents the value of the 2nd basic variable i.e., x_{21}^2,
$5-\theta$ represents the value of the 3rd variable i.e., x_{22}^2,
$7+0\theta$ represents the value of the 4th basic variable i.e., s_1^2,
$9+0\theta$ represents the value of the 5th basic variable i.e., s_2^2 and.
$5+0\theta$ represents the value of the 6th variable i.e., s_4^2 in a more-for-less solution of crisp transportation problem with mixed constraints represented by Table 3.4.

For example, substituting $\theta = 5$, the obtained values of basic variables in a more-for-less solution are

$$x_{12}^2 = 10,$$

$$x_{21}^2 = 13,$$

$$x_{22}^2 = 0,$$

$$s_1^2 = 7,$$

$$s_2^2 = 9,$$

$s_4^2 = 5.$

(ii) For any value θ such that $0 < \theta \leq 5$,
$5 + \theta$ represents the value of the 1st basic variable i.e., x_{12}^2,
$8 + 0\theta$ represents the value of the 2nd basic variable i.e., x_{21}^2,
$5 - \theta$ represents the value of the 3rd variable i.e., x_{22}^2,
$7 - \theta$ represents the value of the 4th basic variable i.e., s_1^2,
$9 + 0\theta$ represents the value of the 5th basic variable i.e., s_2^2 and.
$5 + \theta$ represents the value of the 6th variable i.e., s_4^2 in a more-for-less solution of crisp transportation problem with mixed constraints represented by Table 3.4.

For example, substituting $\theta = 5$, the obtained values of basic variables in a more-for-less solution are

$$x_{12}^2 = 10,$$

$$x_{21}^2 = 8,$$

$$x_{22}^2 = 0,$$

$$s_1^2 = 2,$$

$$s_2^2 = 9,$$

$$s_4^2 = 10.$$

3.8.2 *All More-For-Less Solutions of the Second Problem*

Using the proposed Mehar method-II, all more-for-less solutions (if exists) of the existing symmetric triangular fuzzy transportation problem with mixed constraints [6], represented by Table 3.5, can be obtained as follows.

Table 3.5 Triangular fuzzy transportation problem with mixed constraints [6]

	D_1	D_2	D_3	Availability
S_1	$\tilde{c}_{11}$	$\tilde{c}_{12}$	$\tilde{c}_{13}$	$=_{Mag}\tilde{a}_1$
S_2	$\tilde{c}_{21}$	$\tilde{c}_{22}$	$\tilde{c}_{23}$	$\geq_{Mag}\tilde{a}_2$
S_3	$\tilde{c}_{31}$	$\tilde{c}_{32}$	$\tilde{c}_{33}$	$\leq_{Mag}\tilde{a}_3$
Demand	$=_{Mag}\tilde{b}_1$	$\geq_{Mag}\tilde{b}_2$	$\leq_{Mag}\tilde{b}_3$	

where,

(i) $\tilde{c}_{11} = (2, 1(1-\alpha), 1(1-\alpha)), \tilde{c}_{12} = (5, 3(1-\alpha), 3(1-\alpha)), \tilde{c}_{13} = (4, 2(1-\alpha), 2(1-\alpha)), \tilde{c}_{21} = (6, 4(1-\alpha), 4(1-\alpha)), \tilde{c}_{22} = (3, 2(1-\alpha), 2(1-\alpha)), \tilde{c}_{23} = (1, 1(1-\alpha), 1(1-\alpha)), \tilde{c}_{31} = (8, 4(1-\alpha), 4(1-\alpha)), \tilde{c}_{32} = (9, 6(1-\alpha), 6(1-\alpha)), \tilde{c}_{33} = (2, 1(1-\alpha), 1(1-\alpha))$.

(ii) $\tilde{a}_1 = (5, 3(1-\alpha), 3(1-\alpha)), \tilde{a}_2 = (6, 3(1-\alpha), 3(1-\alpha)), \tilde{a}_3 = (9, 6(1-\alpha), 6(1-\alpha))$.

(iii) $\tilde{b}_1 = (8, 4(1-\alpha), 4(1-\alpha)), \tilde{b}_2 = (10, 2(1-\alpha), 2(1-\alpha)), \tilde{b}_3 = (5, 2(1-\alpha), 2(1-\alpha))$.

Step 1: According to **Step 1** of the proposed Mehar method-II, there is a need to use **Step 1.1** to **Step 1.4** to check that for the symmetric triangular fuzzy transportation problem with mixed constraints, represented by Table 3.5, at least one more-for-less solution will exist or not.

Step 1.1: According to **Step 1.1** of the proposed Mehar method-II, the symmetric triangular fuzzy transportation problem with mixed constraints, represented by Table 3.5, can be transformed into its equivalent crisp transportation problem with mixed constraints represented by Table 3.6.

Step 1.2: According to **Step 1.2** of the proposed Mehar method-II, there is a need to find an optimal solution of the crisp linear programming problem (*P*3.12).

Problem (*P*3.12)

$Maximize\left(5u_1^2 + 6u_2^2 + 9u_3^2 + 8v_1^2 + 10v_2^2 + 5v_3^2\right)$

Subject to

$$u_1^2 + v_1^2 \leq 2,$$

$$u_1^2 + v_2^2 \leq 5,$$

$$u_1^2 + v_3^2 \leq 4,$$

$$u_2^2 + v_1^2 \leq 6,$$

$$u_2^2 + v_2^2 \leq 3,$$

$$u_2^2 + v_3^2 \leq 1,$$

$$u_3^2 + v_1^2 \leq 8,$$

$$u_3^2 + v_2^2 \le 9,$$

$$u_3^2 + v_3^2 \le 2,$$

$$u_1^2, v_1^2 \text{ are unrestricted in sign},$$

$$u_2^2, v_2^2 \ge 0,$$

$$u_3^2, v_3^2 \le 0.$$

It can be easily verified that on solving the problem ($P3.12$), the obtained optimal solution is $u_1^2 = -4$, $u_2^2 = 0$, $u_3^2 = 0$, $v_1^2 = 6$, $v_2^2 = 3$, $v_3^2 = 0$.
Step 1.3: According to **Step 1.3** of the proposed Mehar method-II, the following shadow price matrix is obtained.

$$\begin{bmatrix} u_1^2 + v_1^2 = -1 & u_1^2 + v_2^2 = 1 & u_1^2 + v_3^2 = -6 \\ u_2^2 + v_1^2 = 5 & u_2^2 + v_2^2 = 7 & u_2^2 + v_3^2 = 0 \\ u_3^2 + v_1^2 = 5 & u_3^2 + v_2^2 = 7 & u_3^2 + v_3^2 = 0 \end{bmatrix}$$

Step 1.4: Since, some elements of the shadow price matrix, obtained in **Step 1.3**, are negative. So, according to Case (ii) of **Step 1.4** of the proposed Mehar method-II, there will exist at least one more-for-less solution of symmetric triangular fuzzy transportation problem with mixed constraints represented by Table 3.5.

Step 2: According to Step 2 of the proposed Mehar method-II, all more-for-less solutions of the symmetric triangular fuzzy transportation problem with mixed constraints, represented by Table 3.5, can be obtained as follows.

Step 2.1: It is obvious from **Step 1.3** that $u_1^2 + v_2^2 < 0$ and $u_1^2 + v_3^2 < 0$. Therefore, according to **Step 2.1** of the proposed Mehar method-II,

(i) $R_{12} = \begin{bmatrix} 5+\theta \\ 6 \\ 9 \\ 8 \\ 10+\theta \\ 5 \end{bmatrix}$

(ii) $R_{13} = \begin{bmatrix} 5+\theta \\ 6 \\ 9 \\ 8 \\ 10 \\ 5+\theta \end{bmatrix}$

Step 2.2: According to **Step 2.2** of the proposed Mehar method-II, there is a need to identify the basic variables in an optimal solution of the crisp linear programming problem ($P3.13$).

Problem ($P3.13$) $Minimize \begin{pmatrix} Z = 2x_{11}^2 + 5x_{12}^2 \\ +4x_{13}^2 + 6x_{21}^2 \\ +3x_{22}^2 + x_{23}^2 + 8x_{31}^2 \\ +9x_{32}^2 + 2x_{33}^2 \end{pmatrix}$

Subject to

$$x_{11}^2 + x_{12}^2 + x_{13}^2 = 5,$$

$$x_{21}^2 + x_{22}^2 + x_{23}^2 - s_1^2 = 6,$$

$$x_{31}^2 + x_{32}^2 + x_{33}^2 + s_2^2 = 9,$$

$$x_{11}^2 + x_{21}^2 + x_{31}^2 = 8,$$

$$x_{12}^2 + x_{22}^2 + x_{32}^2 - s_3^2 = 10,$$

$$x_{13}^2 + x_{23}^2 + x_{33}^2 + s_4^2 = 5,$$

$x_{ij}^2 \geq 0; i = 1, 2, 3; j = 1, 2, 3; s_k^2 \geq 0; k = 1, 2, 3, 4.$

It can be easily verified that the following optimal solution is obtained on solving the crisp linear programming Problem ($P3.13$).

$x_{11}^2 = 5, x_{12}^2 = 0, x_{13}^2 = 0, x_{21}^2 = 3, x_{22}^2 = 10, x_{23}^2 = 0, x_{31}^2 = 0, x_{32}^2 = 0, x_{33}^2 = 0, s_1^2 = 7, s_2^2 = 9, s_3^2 = 0, s_4^2 = 5.$

Also, it can be easily verified that the basic variables in this obtained optimal solution are $x_{11}^2, x_{21}^2, x_{22}^2, s_1^2, s_2^2$ and s_4^2.

Step 2.3: According to **Step 2.3** of the proposed Mehar method-II, there is a need to construct the basis matrix B with the help of the coefficients of optimal basic variables $x_{11}^2, x_{21}^2, x_{22}^2, s_1^2, s_2^2, s_4^2$ in the constraints of the crisp linear programming problem ($P3.13$).

$$B = \begin{bmatrix} 1 & 0 & 0 & 0 & 0 & 0 \\ 0 & 1 & 1 & -1 & 0 & 0 \\ 0 & 0 & 0 & 0 & 1 & 0 \\ 1 & 1 & 0 & 0 & 0 & 0 \\ 0 & 0 & 1 & 0 & 0 & 0 \\ 0 & 0 & 0 & 0 & 0 & 1 \end{bmatrix}$$

Step 2.4: According to **Step 2.4** of the proposed Mehar method-II, there is a need to find the multiplicative inverse of the basis matrix B.

$$B^{-1} = \begin{bmatrix} 1 & 0 & 0 & 0 & 0 & 0 \\ -1 & 0 & 0 & 1 & 0 & 0 \\ 0 & 0 & 0 & 0 & 1 & 0 \\ -1 & -1 & 0 & 1 & 1 & 0 \\ 0 & 0 & 1 & 0 & 0 & 0 \\ 0 & 0 & 0 & 0 & 0 & 1 \end{bmatrix}$$

Step 2.5: According to **Step 2.5** of the proposed Mehar method-II, there is need to find $B^{-1}R_{12}$ and $B^{-1}R_{13}$.

(i) $$B^{-1}R_{12} = \begin{bmatrix} 1 & 0 & 0 & 0 & 0 & 0 \\ -1 & 0 & 0 & 1 & 0 & 0 \\ 0 & 0 & 0 & 0 & 1 & 0 \\ -1 & -1 & 0 & 1 & 1 & 0 \\ 0 & 0 & 1 & 0 & 0 & 0 \\ 0 & 0 & 0 & 0 & 0 & 1 \end{bmatrix} \begin{bmatrix} 5+\theta \\ 6 \\ 9 \\ 8 \\ 10+\theta \\ 5 \end{bmatrix} = \begin{bmatrix} 5+\theta \\ 3-\theta \\ 10+\theta \\ 7+0\theta \\ 9+0\theta \\ 5+0\theta \end{bmatrix}$$

(ii) $$B^{-1}R_{13} = \begin{bmatrix} 1 & 0 & 0 & 0 & 0 & 0 \\ -1 & 0 & 0 & 1 & 0 & 0 \\ 0 & 0 & 0 & 0 & 1 & 0 \\ -1 & -1 & 0 & 1 & 1 & 0 \\ 0 & 0 & 1 & 0 & 0 & 0 \\ 0 & 0 & 0 & 0 & 0 & 1 \end{bmatrix} \begin{bmatrix} 5+\theta \\ 6 \\ 9 \\ 8 \\ 10 \\ 5+\theta \end{bmatrix} = \begin{bmatrix} 5+\theta \\ 3-\theta \\ 10+0\theta \\ 7-\theta \\ 9+0\theta \\ 5+\theta \end{bmatrix}$$

Step 2.6: According to **Step 2.6** of the proposed Mehar method-II, there is a need to find the range of θ (where, $\theta > 0$) so that all the elements of the matrix, obtained in **Step 2.5**, are greater than or equal to zero.

It is obvious from **Step 2.5** that

(i) The range of θ for which all the elements of the matrix $B^{-1}R_{12}$, obtained in **Step 2.5**, are greater than or equal to zero, i.e., $5+\theta \geq 0, 3-\theta \geq 0, 10+\theta \geq 0$ is $0 < \theta \leq 3$.

(ii) The range of θ for which all the elements of the matrix $B^{-1}R_{13}$, obtained in **Step 2.5**, are greater than or equal to zero, i.e., $5+\theta \geq 0, 3-\theta \geq 0, 7-\theta \geq 0$ is $0 < \theta \leq 3$.

Step 2.7: According to **Step 2.7** of the proposed Mehar method-II,

(i) For any value θ such that $0 < \theta \leq 3$,

$5+\theta$ represents the value of the 1st basic variable i.e., x_{11}^2,
$3-\theta$ represents the value of the 2nd basic variable i.e., x_{21}^2,
$10+\theta$ represents the value of the 3rd variable i.e., x_{22}^2,
$7+0\theta$ represents the value of the 4th basic variable i.e., s_1^2,
$9+0\theta$ represents the value of the 5th basic variable i.e., s_2^2 and.
$5+0\theta$ represents the value of the 6th variable i.e., s_4^2 in a more-for-less solution of crisp transportation problem with mixed constraints represented by Table 3.6.

For example, substituting $\theta = 3$, the obtained values of basic variables in a more-for-less solution are

$$x_{11}^2 = 8,$$

$$x_{21}^2 = 0,$$

$$x_{22}^2 = 13,$$

$$s_1^2 = 7,$$

$$s_2^2 = 9,$$

$$s_4^2 = 5$$

(ii) For any value θ such that $0 < \theta \leq 3$,
$5+\theta$ represents the value of the 1st basic variable i.e., x_{11}^2,
$3-\theta$ represents the value of the 2nd basic variable i.e., x_{21}^2,
$10+0\theta$ represents the value of the 3rd variable i.e., x_{22}^2,
$7-\theta$ represents the value of the 4th basic variable i.e., s_1^2,
$9+0\theta$ represents the value of the 5th basic variable i.e., s_2^2 and.
$5+\theta$ represents the value of the 6th variable i.e., s_4^2 in a more-for-less solution of crisp transportation problem with mixed constraints represented by Table 3.6.

For example, substituting $\theta = 3$, the obtained values of basic variables in a more-for-less solution are

$$x_{11}^2 = 8,$$

$$x_{21}^2 = 0,$$

$$x_{22}^2 = 10,$$

$$s_1^2 = 4,$$

$$s_2^2 = 9,$$

$$s_4^2 = 8.$$

Step 2.8: According to **Step 2.8** of the proposed Mehar method-II,

(i) For any value θ such that $0 < \theta \leq 3$,

$(5+\theta, \delta_{114}(1-\alpha), \delta_{114}(1-\alpha))$ represents the value of the 1st basic variable i.e., $\tilde{x}_{11}$,

$(3-\theta, \delta_{214}(1-\alpha), \delta_{214}(1-\alpha))$ represents the value of the 2nd basic variable i.e., $\tilde{x}_{21}$,

$(10+\theta, \delta_{224}(1-\alpha), \delta_{224}(1-\alpha))$ represents the value of the 3rd variable i.e., $\tilde{x}_{22}$,

$(7+0\theta, \gamma_{11}(1-\alpha), \gamma_{11}(1-\alpha))$ represents the value of the 4th basic variable i.e., $\tilde{s}_1$,

$(9+0\theta, \gamma_{21}(1-\alpha), \gamma_{21}(1-\alpha))$ represents the value of the 5th basic variable i.e., $\tilde{s}_2$ and $(5+0\theta, \gamma_{41}(1-\alpha), \gamma_{41}(1-\alpha))$ represents the value of the 6th variable i.e., $\tilde{s}_4$ in a more-for-less solution of the symmetric triangular fuzzy transportation problem with mixed constraints represented by Table 3.5.

For example,

(a) Substituting $\theta = 3$, $\delta_{ij4} = 2 \forall i, j, \gamma_{k1} = 1; k = 1, 2, 4$, the obtained values of basic variables in a more-for-less solution are

$$\tilde{x}_{11} =_{Mag} (8, 2(1-\alpha), 2(1-\alpha)),$$

$$\tilde{x}_{21} =_{Mag} (0, 2(1-\alpha), 2(1-\alpha)),$$

$$\tilde{x}_{22} =_{Mag} (13, 2(1-\alpha), 2(1-\alpha)),$$

$$\tilde{s}_1 =_{Mag} (7, 1(1-\alpha), 1(1-\alpha)),$$

$$\tilde{s}_2 =_{Mag} (9, 1(1-\alpha), 1(1-\alpha)),$$

$\tilde{s}_4 =_{Mag} (5, 1(1-\alpha), 1(1-\alpha))$ and the corresponding symmetic triangular fuzzy transportation cost is $\tilde{Z} =_{Mag} (55, 2(1-\alpha), 2(1-\alpha))$.

(b) Substituting $\theta = 3$, $\delta_{ij4} = 4 \forall i, j, \gamma_{k1} = 1; k = 1, 2, 4$, the obtained values of basic variables in a more-for-less solution are

$$\tilde{x}_{11} =_{Mag} (8, 4(1-\alpha), 4(1-\alpha)),$$

$$\tilde{x}_{21} =_{Mag} (0, 4(1-\alpha), 4(1-\alpha)),$$

$$\tilde{x}_{22} =_{Mag} (13, 4(1-\alpha), 4(1-\alpha)),$$

$$\tilde{s}_1 =_{Mag} (7, 1(1-\alpha), 1(1-\alpha)),$$

$\tilde{s}_2 =_{Mag} (9, 1(1-\alpha), 1(1-\alpha))$, $\tilde{s}_4 =_{Mag} (5, 1(1-\alpha), 1(1-\alpha))$ and the corresponding symmetric triangular fuzzy transportation cost is $\tilde{Z} =_{Mag} (55, 4(1-\alpha), 4(1-\alpha))$

(ii) For any value θ such that $0 < \theta \le 3$,

$(5+\theta, \delta_{114}(1-\alpha), \delta_{114}(1-\alpha))$ represents the value of the 1st basic variable i.e., $\tilde{x}_{11}$,

$(3-\theta, \delta_{214}(1-\alpha), \delta_{214}(1-\alpha))$ represents the value of the 2nd basic variable i.e., $\tilde{x}_{21}$,

$(10+0\theta, \delta_{224}(1-\alpha), \delta_{224}(1-\alpha))$ represents the value of the 3rd variable i.e., $\tilde{x}_{22}$,

$(7-\theta, \gamma_{11}(1-\alpha), \gamma_{11}(1-\alpha))$ represents the value of the 4th basic variable i.e., $\tilde{s}_1$,

$(9+0\theta, \gamma_{21}(1-\alpha), \gamma_{21}(1-\alpha))$ represents the value of the 5th basic variable i.e., $\tilde{s}_2$ and.

$(5+\theta, \gamma_{41}(1-\alpha), \gamma_{41}(1-\alpha))$ represents the value of the 6th variable i.e., $\tilde{s}_4$ in a more-for-less solution of the symmetric triangular fuzzy transportation problem with mixed constraints represented by Table 3.5.

For example,

(a) Substituting $\theta = 3$, $\delta_{ij4} = 2 \forall i, j$, $\gamma_{k1} = 1; k = 1, 2, 4$, the obtained values of basic variables in a more-for-less solution are

$$\tilde{x}_{11} =_{Mag} (8, 2(1-\alpha), 2(1-\alpha)),$$

$$\tilde{x}_{21} =_{Mag} (0, 2(1-\alpha), 2(1-\alpha)),$$

$$\tilde{x}_{22} =_{Mag} (10, 2(1-\alpha), 2(1-\alpha)),$$

$$\tilde{s}_1 =_{Mag} (4, 1(1-\alpha), 1(1-\alpha)),$$

$$\tilde{s}_2 =_{Mag} (9, 1(1-\alpha), 1(1-\alpha)),$$

$\tilde{s}_4 =_{Mag} (8, 1(1-\alpha), 1(1-\alpha))$ and the corresponding symmetric triangular fuzzy transportation cost is $\tilde{Z} =_{Mag} (46, 2(1-\alpha), 2(1-\alpha))$.

(b) Substituting $\theta = 3$, $\delta_{ij4} = 3 \forall i, j$, $\gamma_{k1} = 1; k = 1, 2, 4$, the obtained values of basic variables in a more-for-less solution are

$$\tilde{x}_{11} =_{Mag} (8, 3(1-\alpha), 3(1-\alpha)),$$

$$\tilde{x}_{21} =_{Mag} (0, 3(1-\alpha), 3(1-\alpha)),$$

$$\tilde{x}_{22} =_{Mag} (10, 3(1-\alpha), 3(1-\alpha)),$$

$$\tilde{s}_{1} =_{Mag} (4, 1(1-\alpha), 1(1-\alpha)),$$

$$\tilde{s}_{2} =_{Mag} (9, 1(1-\alpha), 1(1-\alpha)),$$

$\tilde{s}_4 =_{Mag} (8, 1(1-\alpha), 1(1-\alpha))$ and the corresponding symmetric triangular fuzzy transportation cost is $\tilde{Z} =_{Mag} (46, 3(1-\alpha), 3(1-\alpha))$

3.9 Results and Discussion

The results of the existing crisp transportation problem with mixed constraints [2], represented by Table 3.4 and the existing symmetric triangular fuzzy transportation problem with mixed constraints [6], represented by Table 3.5, obtained by the existing methods [2, 6] and the proposed Mehar method-II, are shown in Table 3.7.

It is obvious from Table 3.7 that.

(i) If the optimal values of the dual variables, obtained by the existing methods [2, 6], are substituted in the objective function i.e., $5u_1^2+6u_2^2+9u_3^2+8v_1^2+10v_2^2+5v_3^2$ of the crisp linear programming problem ($P3.10$) corresponding to the first problem [2]. Then, the obtained total transportation cost is 94. While, if the optimal values of the dual variables, obtained by the proposed Mehar method-II, are substituted in the objective function i.e., $5u_1^2+6u_2^2+9u_3^2+8v_1^2+10v_2^2+5v_3^2$ of the crisp linear programming problem ($P3.10$) corresponding to the first problem [2]. Then, the obtained total transportation cost is 80.

(ii) If the optimal values of the dual variables, obtained by the existing method [6], are substituted in the objective function i.e., $5u_1^2+6u_2^2+9u_3^2+8v_1^2+10v_2^2+5v_3^2$ of the crisp linear programming problem ($P3.12$) corresponding to the second problem [6]. Then, the obtained total transportation cost is 72. While, if the optimal values of the dual variables, obtained by the proposed Mehar method-II, are substituted in the objective function i.e., $5u_1^2 + 6u_2^2 + 9u_3^2 + 8v_1^2 + 10v_2^2 + 5v_3^2$ of the crisp linear programming problem ($P3.12$) corresponding to the second problem [6]. Then, the obtained total transportation cost is 58. Also, it is obvious that the existing method [2] cannot be used to solve the second problem as the existing method [2] is proposed by considering the assumption

Table 3.6 Crisp transportation problem with mixed constraints

	D_1	D_2	D_3	Availability
S_1	2	5	4	$= 5$
S_2	6	3	1	≥ 6
S_3	8	9	2	≤ 9
Demand	$= 8$	≥ 10	≤ 5	

Table 3.7 Results of the existing problems

	Optimal values of u_i^2 and v_j^2 and corresponding total transportation cost	
	First problem [2]	Second problem [6]
Adlakha et al.'s method [2]	$u_1^2 = -6, u_2^2 = 0, u_3^2 = 1, v_1^2 = 5, v_2^2 = 7, v_3^2 = 1$ and corresponding total transportation cost is 94	Not applicable
Vidhya et al.'s method [6]	$u_1^2 = -6, u_2^2 = 0, u_3^2 = 1, v_1^2 = 5, v_2^2 = 7, v_3^2 = 1$ and corresponding total transportation cost is 94	$u_1^2 = Mag(\tilde{u}_1) = -4, u_2^2 = Mag(\tilde{u}_2) = 0, u_3^2 = Mag(\tilde{u}_3) = 1, v_1^2 = Mag(\tilde{v}_1) = 6, v_2^2 = Mag(\tilde{v}_2) = 3, v_3^2 = Mag(\tilde{v}_3) = 1$ and corresponding total transportation cost is 72
Proposed Mehar method-II	$u_1^2 = -6, u_2^2 = 0, u_3^2 = 0, v_1^2 = 5, v_2^2 = 7, v_3^2 = 0$ and corresponding total transportation cost is 80	$u_1^2 = Mag(\tilde{u}_1) = -4, u_2^2 = Mag(\tilde{u}_2) = 0, u_3^2 = Mag(\tilde{u}_3) = 0, v_1^2 = Mag(\tilde{v}_1) = 6, v_2^2 = Mag(\tilde{v}_2) = 3, v_3^2 = Mag(\tilde{v}_3) = 0$ and corresponding total transportation cost is 58

that each parameter is a positive real number. While, each parameter of the second problem is a symmetric triangular fuzzy number.

Although, it is obvious that for both the problems the total transportation cost, obtained by the proposed Mehar method-II, is better than the total transportation cost, obtained by the existing methods [2, 6]. However, the following questions naturally arise in mind.

First question: Are the values of u_i^2 and v_j^2, obtained by the existing methods [2, 6], correct?

Second question: Are the values of u_i^2 and v_j^2, obtained by the proposed Mehar method-II, correct?

Third question: Why the values of u_i^2 and v_j^2, obtained by the existing methods [2, 6], are different from the proposed Mehar method-II?

Fourth question: What is the impact of incorrect values of u_i^2 and v_j^2 on more-for-less solution?

3.9.1 Response of the First Question

The following clearly indicates that the values of u_i^2 and v_j^2, obtained by the existing methods [2, 6], are not correct.

(i) For the first problem, the values of u_i^2 and v_j^2, obtained by the existing methods [2, 6], represents the optimal decision variables of the dual problem ($P3.10$) corresponding to the crisp linear programming problem ($P3.11$). If these values are correct. Then, according to strong duality theorem [5], the value, obtained on substituting the obtained values of u_i^2 and v_j^2 in the objective function i.e., $5u_1^2 + 6u_2^2 + 9u_3^2 + 8v_1^2 + 10v_2^2 + 5v_3^2$ of the dual problem ($P3.10$), should be same as the optimal value of the crisp linear programming problem ($P3.11$). However, as discussed in Sect. 3.8.1 that the optimal value of the crisp linear programming problem ($P3.11$) is 80. While, it is obvious from Table 3.7 that the value, obtained on substituting the obtained values of u_i^2 and v_j^2 in the objective function i.e., $5u_1^2 + 6u_2^2 + 9u_3^2 + 8v_1^2 + 10v_2^2 + 5v_3^2$ of the dual problem ($P3.10$) is 94.

Also, it is pertinent to mention that

(a) As the sign of 3rd constraint of the crisp linear programming problem ($P3.11$) is $\leq$. So, according to duality theory [5], the dual variable u_3^2 should be ≤ 0 i.e., $u_3^2 \leq 0$. While, it is obvious that for the obtained value of u_3^2 this condition is not satisfying.

(b) As the sign of 6th constraint of the crisp linear programming problem ($P3.11$) is $\leq$. So, according to duality theory [5], the dual variable v_3^2

should be ≤ 0 i.e., $v_3^2 \leq 0$. While, it is obvious that for the obtained value of v_3^2 this condition is not satisfying.

(ii) For the second problem, the values of u_i^2 and v_j^2, obtained by the existing methods [2, 6], represents the optimal decision variables of the dual problem ($P3.12$) corresponding to the crisp linear programming problem ($P3.13$). If these values are correct. Then, according to strong duality theorem [5], the value, obtained on substituting the obtained values of u_i^2 and v_j^2 in the objective function i.e., $5u_1^2 + 6u_2^2 + 9u_3^2 + 8v_1^2 + 10v_2^2 + 5v_3^2$ of the dual problem ($P3.12$), should be same as the optimal value of the crisp linear programming problem ($P3.13$). However, as discussed in Sect. 3.8.2 that the optimal value of the crisp linear programming problem ($P3.13$) is 58. While, it is obvious from Table 3.7 that the value, obtained on substituting the obtained values of u_i^2 and v_j^2 in the objective function i.e., $5u_1^2 + 6u_2^2 + 9u_3^2 + 8v_1^2 + 10v_2^2 + 5v_3^2$ of the dual problem ($P3.12$) is 72.

Also, it is pertinent to mention that

(a) As the sign of 3rd constraint of the crisp linear programming problem ($P3.13$) is $\leq$. So, according to duality theory [5], the dual variable u_3^2 should be ≤ 0 i.e., $u_3^2 \leq 0$. While, it is obvious that for the obtained value of u_3^2 this condition is not satisfying.
(b) As the sign of 6th constraint of the crisp linear programming problem ($P3.13$) is $\leq$. So, according to duality theory [5], the dual variable v_3^2 should be ≤ 0 i.e., $v_3^2 \leq 0$. While, it is obvious that for the obtained value of v_3^2 this condition is not satisfying.

3.9.2 *Response of the Second Question*

The following clearly indicates that the values of u_i^2 and v_j^2, obtained by the proposed Mehar method-II, are correct.

(i) For the first problem, the values of u_i^2 and v_j^2, , obtained by the proposed Mehar method-II, represents the optimal decision variables of the dual problem ($P3.10$) corresponding to the crisp linear programming problem ($P3.11$). If these values are correct. Then, according to strong duality theorem [5], the value, obtained on substituting the obtained values of u_i^2 and v_j^2 in the objective function i.e., $5u_1^2 + 6u_2^2 + 9u_3^2 + 8v_1^2 + 10v_2^2 + 5v_3^2$ of the dual problem ($P3.10$), should be same as the optimal value of the crisp linear programming problem ($P3.11$). It is obvious from Sect. 3.8.1 that the optimal value of the crisp linear programming problem ($P3.11$) is 80. Also, it is obvious from Table 3.7 that the value, obtained on substituting the obtained values of u_i^2 and v_j^2 in the objective function i.e., $5u_1^2 + 6u_2^2 + 9u_3^2 + 8v_1^2 + 10v_2^2 + 5v_3^2$ of the dual problem ($P3.10$) is 80.

Furthermore, it is pertinent to mention that

(a) As the sign of 3rd constraint of the crisp linear programming problem ($P3.11$) is $\leq$. So, according to duality theory [5], the dual variable u_3^2 should be ≤ 0 i.e., $u_3^2 \leq 0$. It is obvious from Table 3.7 that for the obtained value of u_3^2 this condition is satisfying.
(b) As the sign of 6th constraint of the crisp linear programming problem ($P3.11$) is $\leq$. So, according to duality theory [5], the dual variable v_3^2 should be ≤ 0 i.e., $v_3^2 \leq 0$. It is obvious from Table 3.7 that for the obtained value of v_3^2 this condition is satisfying.

(ii) For the second problem, the values of u_i^2 and v_j^2, obtained by the proposed Mehar method-II, represents the optimal decision variables of the dual problem ($P3.12$) corresponding to the crisp linear programming problem ($P3.13$). If these values are correct. Then, according to strong duality theorem [5], the value, obtained on substituting the obtained values of u_i^2 and v_j^2 in the objective function i.e., $5u_1^2 + 6u_2^2 + 9u_3^2 + 8v_1^2 + 10v_2^2 + 5v_3^2$ of the dual problem ($P3.12$), should be same as the optimal value of the crisp linear programming problem ($P3.13$). It is obvious from Sect. 3.8.2 that the optimal value of the crisp linear programming problem ($P3.13$) is 58. Also, it is obvious from Table 3.7 that the value, obtained on substituting the obtained values of u_i^2 and v_j^2 in the objective function i.e., $5u_1^2 + 6u_2^2 + 9u_3^2 + 8v_1^2 + 10v_2^2 + 5v_3^2$ of the dual problem ($P3.12$) is 58
Furthermore, it is pertinent to mention that

(a) As the sign of 3rd constraint of the crisp linear programming problem ($P3.13$) is $\leq$. So, according to duality theory [5], the dual variable u_3^2 should be ≤ 0 i.e., $u_3^2 \leq 0$. It is obvious from Table 3.7 that for the obtained value of u_3^2 this condition is satisfying.
(b) As the sign of 6th constraint of the crisp linear programming problem ($P3.13$) is $\leq$. So, according to duality theory [5], the dual variable v_3^2 should be ≤ 0 i.e., $v_3^2 \leq 0$. It is obvious from Table 3.7 that for the obtained value of v_3^2 this condition is satisfying.

3.9.3 Response of the Third Question

It is pertinent to mention that.

(i) For the first problem, the values of u_i^2 and v_j^2, obtained by the proposed Mehar method-II, represents an optimal solution of the dual problem ($P3.10$).
(ii) For the second problem, the values of u_i^2 and v_j^2, obtained by the proposed Mehar method-II, represents an optimal solution of the dual problem ($P3.12$).

While, in the existing methods [2, 6], the following procedure is used to find the values of u_i^2 and v_j^2.

It is a well-known fact that the number of basic variables in an optimal solution of a crisp balanced transportation problem having m sources and n destinations will be exactly equal to $(m + n - 1)$. On the same direction, in the existing methods [2, 6], it is assumed that the number of basic variables in an optimal solution of crisp/fuzzy transportation problem with mixed constraints having m sources and n destinations will be exactly equal to $(m + n - 1)$.

While, in actual case, the number of basic variables in an optimal solution of crisp/fuzzy transportation problem with mixed constraints having m sources and n destinations will not necessarily be equal to $(m + n - 1)$.

The following clearly indicates that the above-mentioned assumption is considered in the existing methods [2, 6].

3.9.3.1 Validity of the Claim for Vidhya et al.'s Method

Vidhya et al. [6] considered the symmetric triangular fuzzy transportation problem with mixed constraints, represented by Table 3.5, to illustrate their proposed method.

Vidhya et al. [6] claimed that on solving the symmetric triangular fuzzy transportation problem with mixed constraints, represented by Table 3.5, the following fuzzy optimal solution is obtained.

$$\tilde{x}_{11} =_{Mag} (5, 4(1-\alpha), 4(1-\alpha)),$$

$$\tilde{x}_{12} =_{Mag} (0, 0(1-\alpha), 0(1-\alpha)),$$

$$\tilde{x}_{13} =_{Mag} (0, 0(1-\alpha), 0(1-\alpha)),$$

$$\tilde{x}_{21} =_{Mag} (3, 4(1-\alpha), 4(1-\alpha)),$$

$$\tilde{x}_{22} =_{Mag} (10, 3(1-\alpha), 3(1-\alpha)),$$

$$\tilde{x}_{23} =_{Mag} (0, 3(1-\alpha), 3(1-\alpha)),$$

$$\tilde{x}_{31} =_{Mag} (0, 0(1-\alpha), 0(1-\alpha)),$$

$$\tilde{x}_{32} =_{Mag} (0, 0(1-\alpha), 0(1-\alpha)),$$

$$\tilde{x}_{33} =_{Mag} (0, 6(1-\alpha), 6(1-\alpha)).$$

Vidhya et al. [6] further claimed that it is obvious from the obtained optimal solution that $\tilde{x}_{11}, \tilde{x}_{21}, \tilde{x}_{22}$ are basic variables as the values of these variables are non-zero. However, as there are 3 sources and 3 destinations, so the number of basic variables should be $3 + 3 - 1 = 5$. Hence, Vidhya et al. [6] have considered $\tilde{x}_{23}$ and $\tilde{x}_{33}$ as basic variables.

While, the following clearly indicates that in actual case,

(a) The number of basic variables is 6.
(b) $\tilde{x}_{23}$and $\tilde{x}_{33}$ are not basic variables.

The symmetric triangular fuzzy transportation problem with mixed constraints, represented by Table 3.5, can be transformed into its equivalent fuzzy linear programming problem (P3.14).

Problem (P3.14) *Minimize*$\left(\tilde{Z}\right)$

Subject to

$\tilde{x}_{11} \oplus \tilde{x}_{12} \oplus \tilde{x}_{13} =_{Mag} (5, 3(1-\alpha), 3(1-\alpha))$,
$\tilde{x}_{21} \oplus \tilde{x}_{22} \oplus \tilde{x}_{23} \succeq_{Mag} (6, 3(1-\alpha), 3(1-\alpha))$,
$\tilde{x}_{31} \oplus \tilde{x}_{32} \oplus \tilde{x}_{33} \preceq_{Mag} (9, 6(1-\alpha), 6(1-\alpha))$,
$\tilde{x}_{11} \oplus \tilde{x}_{21} \oplus \tilde{x}_{31} =_{Mag} (8, 4(1-\alpha), 4(1-\alpha))$,
$\tilde{x}_{12} \oplus \tilde{x}_{22} \oplus \tilde{x}_{32} \succeq_{Mag} (10, 2(1-\alpha), 2(1-\alpha))$,
$\tilde{x}_{13} \oplus \tilde{x}_{23} \oplus \tilde{x}_{33} \preceq_{Mag} (5, 2(1-\alpha), 2(1-\alpha))$,
$\tilde{x}_{ij} \succeq_{Mag} \tilde{0};\ i = 1, 2, 3;\ j = 1, 2, 3$,
where,

$$\tilde{z} =_{Mag} \begin{pmatrix} (2, 1(1-\alpha), 1(1-\alpha)) \otimes \tilde{x}_{11} \\ \oplus (5, 3(1-\alpha), 3(1-\alpha)) \otimes \tilde{x}_{12} \\ \oplus (4, 2(1-\alpha), 2(1-\alpha)) \otimes \tilde{x}_{13} \\ \oplus (6, 4(1-\alpha), 4(1-\alpha)) \otimes \tilde{x}_{21} \\ \oplus (3, 2(1-\alpha), 2(1-\alpha)) \otimes \tilde{x}_{22} \\ \oplus (1, 1(1-\alpha), 1(1-\alpha)) \otimes \tilde{x}_{23} \\ \oplus (8, 4(1-\alpha), 4(1-\alpha)) \otimes \tilde{x}_{31} \\ \oplus (9, 6(1-\alpha), 6(1-\alpha)) \otimes \tilde{x}_{23} \\ \oplus (2, 1(1-\alpha), 1(1-\alpha)) \otimes \tilde{x}_{33} \end{pmatrix}.$$

Introducing surplus variables $\tilde{s}_1, \tilde{s}_4$ and slack variables $\tilde{s}_2, \tilde{s}_3$, the fuzzy linear programming problem (P3.14) can be transformed into its equivalent fuzzy linear programming problem (P3.15).

Problem (P3.15) *Minimize* $\left(\tilde{Z}\right)$

Subject to

$\tilde{x}_{11} \oplus \tilde{x}_{12} \oplus \tilde{x}_{13} =_{Mag} (5, 3(1-\alpha), 3(1-\alpha))$,
$\tilde{x}_{21} \oplus \tilde{x}_{22} \oplus \tilde{x}_{23} \ominus \tilde{s}_1 =_{Mag} (6, 3(1-\alpha), 3(1-\alpha))$,
$\tilde{x}_{31} \oplus \tilde{x}_{32} \oplus \tilde{x}_{33} \oplus \tilde{s}_2 =_{Mag} (9, 6(1-\alpha), 6(1-\alpha))$,
$\tilde{x}_{11} \oplus \tilde{x}_{21} \oplus \tilde{x}_{31} =_{Mag} (8, 4(1-\alpha), 4(1-\alpha))$,
$\tilde{x}_{12} \oplus \tilde{x}_{22} \oplus \tilde{x}_{32} \ominus \tilde{s}_3 =_{Mag} (10, 2(1-\alpha), 2(1-\alpha))$,

$\tilde{x}_{13} \oplus \tilde{x}_{23} \oplus \tilde{x}_{33} \oplus \tilde{s}_4 =_{Mag} (5, 2(1-\alpha), 2(1-\alpha)),$
$\tilde{x}_{ij} \geq_{Mag} \tilde{0}; i = 1, 2, 3; j = 1, 2, 3; \tilde{s}_k \geq_{Mag} \tilde{0}; k = 1, 2, 3, 4.$

It can be easily verified that rank of the coefficient matrix of the fuzzy linear programming problem ($P3.15$) i.e, rank of the matrix $\begin{bmatrix} 1\,1\,1\,0\,0\,0\,0\,0\,0 & 0 & 0 & 0 & 0 \\ 0\,0\,0\,1\,1\,1\,0\,0\,0 & -1 & 0 & 0 & 0 \\ 0\,0\,0\,0\,0\,0\,1\,1\,1 & 0 & 1 & 0 & 0 \\ 1\,0\,0\,1\,0\,0\,1\,0\,0 & 0 & 0 & 0 & 0 \\ 0\,1\,0\,0\,1\,0\,0\,1\,0 & 0 & 0 & -1 & 0 \\ 0\,0\,1\,0\,0\,1\,0\,0\,1 & 0 & 0 & 0 & 1 \end{bmatrix}$ is 6.

Hence, the number of basic variables in an optimal solution of the fuzzy transportation problem with mixed constraints, represented by Table 3.5, will be 6.

Also, it can be easily verified that on substituting the non-zero decision variables, in the obtained optimal solution i.e., $\tilde{x}_{11} =_{Mag} (5, 4(1-\alpha), 4(1-\alpha))$, $\tilde{x}_{21} =_{Mag} (3, 4(1-\alpha), 4(1-\alpha))$, $\tilde{x}_{22} =_{Mag} (10, 3(1-\alpha), 3(1-\alpha))$, in the constraints of the fuzzy linear programming problem ($P3.15$), the following values of $\tilde{s}_1$, $\tilde{s}_2$, $\tilde{s}_3$ and $\tilde{s}_4$ are obtained.

$$\tilde{s}_1 =_{Mag} (7, 4(1-\alpha), 4(1-\alpha)),$$

$$\tilde{s}_2 =_{Mag} (9, 6(1-\alpha), 6(1-\alpha)),$$

$$\tilde{s}_3 =_{Mag} (0, 3(1-\alpha), 3(1-\alpha)),$$

$$\tilde{s}_4 =_{Mag} (5, 2(1-\alpha), 2(1-\alpha)).$$

Therefore, the optimal solution (including the values of slack and surplus variables) of the symmetric triangular fuzzy transportation problem with mixed constraints, represented by Table 3.5, is

$$\tilde{x}_{11} =_{Mag} (5, 4(1-\alpha), 4(1-\alpha)),$$

$$\tilde{x}_{12} =_{Mag} (0, 0(1-\alpha), 0(1-\alpha)),$$

$$\tilde{x}_{13} =_{Mag} (0, 0(1-\alpha), 0(1-\alpha)),$$

$$\tilde{x}_{21} =_{Mag} (3, 4(1-\alpha), 4(1-\alpha)),$$

$$\tilde{x}_{22} =_{Mag} (10, 3(1-\alpha), 3(1-\alpha)),$$

$$\tilde{x}_{23} =_{Mag} (0, 3(1-\alpha), 3(1-\alpha)),$$

$$\tilde{x}_{31} =_{Mag} (0, 0(1-\alpha), 0(1-\alpha)),$$

$$\tilde{x}_{32} =_{Mag} (0, 0(1-\alpha), 0(1-\alpha)),$$

$$\tilde{x}_{33} =_{Mag} (0, 6(1-\alpha), 6(1-\alpha)),$$

$$\tilde{s}_1 =_{Mag} (7, 4(1-\alpha), 4(1-\alpha)),$$

$$\tilde{s}_2 =_{Mag} (9, 6(1-\alpha), 6(1-\alpha)),$$

$$\tilde{s}_3 =_{Mag} (0, 3(1-\alpha), 3(1-\alpha)),$$

$$\tilde{s}_4 =_{Mag} (5, 2(1-\alpha), 2(1-\alpha)).$$

Since, the values of $\tilde{s}_1$, $\tilde{s}_2$ and $\tilde{s}_4$ are non-zero. So, the variables $\tilde{x}_{11}, \tilde{x}_{21}, \tilde{x}_{22}, \tilde{s}_1, \tilde{s}_2$ and $\tilde{s}_4$ are basic variables. Hence, $\tilde{x}_{23}$ and $\tilde{x}_{33}$ are not basic variables.

3.9.3.2 Validity of the Claim for Adlakha et al.'s Method

Adlakha et al. [2] considered the crisp transportation problem with mixed constraints, represented by Table 3.4, to illustrate their proposed method.

Adlakha et al. [2] claimed that on solving the crisp transportation problem with mixed constraints, represented by Table 3.4, the following optimal solution is obtained.

$x_{11}^2 = 0, x_{12}^2 = 5, x_{13}^2 = 0, x_{21}^2 = 8, x_{22}^2 = 5, x_{23}^2 = 0, x_{31}^2 = 0, x_{32}^2 = 0, x_{33}^2 = 0.$

Adlakha et al. [2] further claimed that it is obvious from the obtained optimal solution that $x_{12}^2, x_{21}^2, x_{22}^2$ are basic variables as the values of these variables are non-zero. However, as there are 3 sources and 3 destinations, so the number of basic variables should be $3 + 3 - 1 = 5$. Hence, Adlakha et al. [2] have considered x_{23}^2 and x_{33}^2 as basic variables.

While, the following clearly indicates that in actual case,

(a) The number of basic variables is 6.
(b) x_{23}^2 and x_{33}^2 are not basic variables.

The crisp transportation problem with mixed constraints, represented by Table 3.4, can be transformed into its equivalent crisp linear programming problem ($P3.16$).

Problem (*P*3.16)

$$Minimize\left(Z = 10x_{11}^2 + x_{12}^2 + 4x_{13}^2 + 5x_{21}^2 + 7x_{22}^2 + x_{23}^2 + 8x_{31}^2 + 9x_{32}^2 + 2x_{33}^2\right)$$

Subject to

$$x_{11}^2 + x_{12}^2 + x_{13}^2 = 5,$$

$$x_{21}^2 + x_{22}^2 + x_{23}^2 \leq 6,$$

$$x_{31}^2 + x_{32}^2 + x_{33}^2 \geq 9,$$

$$x_{11}^2 + x_{21}^2 + x_{31}^2 = 8,$$

$$x_{12}^2 + x_{22}^2 + x_{32}^2 \geq 10,$$

$$x_{13}^2 + x_{23}^2 + x_{33}^2 \leq 5,$$

$$x_{ij}^2 \geq 0;\ i = 1, 2, 3;\ j = 1, 2, 3.$$

Introducing surplus variables s_1^2, s_4^2 and slack variables s_2^2, s_3^2, the crisp linear programming problem (*P*3.16) can be transformed into its equivalent crisp linear programming problem (*P*3.11).

It can be easily verified that rank of the coefficient matrix of the crisp linear programming problem i.e. (*P*3.11)., rank of the matrix $\begin{bmatrix} 1&1&1&0&0&0&0&0&0&0&0&0&0 \\ 0&0&0&1&1&1&0&0&0&-1&0&0&0 \\ 0&0&0&0&0&0&1&1&1&0&1&0&0 \\ 1&0&0&1&0&0&1&0&0&0&0&0&0 \\ 0&1&0&0&1&0&0&1&0&0&0&-1&0 \\ 0&0&1&0&0&1&0&0&1&0&0&0&1 \end{bmatrix}$ is 6.

Hence, the number of basic variables in an optimal solution of the crisp transportation problem with mixed constraints, represented by Table 3.4, will be 6.

Also, it can be easily verified that on substituting the non-zero decision variables, in the obtained optimal solution i.e., $x_{12}^2 = 5, x_{21}^2 = 8, x_{22}^2 = 5$, in the constraints of the crisp linear programming problem (*P*3.11), the following values of s_1^2, s_2^2, s_3^2 and s_4^2 are obtained.

$s_1^2 = 7, s_2^2 = 9, s_3^2 = 0, s_4^2 = 5.$

Therefore, the optimal solution (including the values of slack and surplus variables) of the crisp transportation problem with mixed constraints, represented by Table 3.4, is.

$x_{11}^2 = 0, x_{12}^2 = 5, x_{13}^2 = 0, x_{21}^2 = 8, x_{22}^2 = 5, x_{23}^2 = 0, x_{31}^2 = 0, x_{32}^2 = 0, x_{33}^2 = 0, s_1^2 = 7, s_2^2 = 9, s_3^2 = 0, s_4^2 = 5$.

Since, the values of s_1^2, s_2^2 and s_4^2 are non-zero. So, the variables $x_{12}^2, x_{21}^2, x_{22}^2, s_1^2, s_2^2$ and s_4^2 are basic variables. Hence, x_{23}^2 and x_{33}^2 are not basic variables.

3.9.4 Response of the Fourth Question

It is pertinent to mention that according to the existing methods [2, 6], at least one more-for-less solution of the crisp/fuzzy transportation problems with mixed constraints, represented by Tables 3.1 and 3.2, will exist if the condition $u_i^2 + v_j^2 < 0$ will be satisfied for at least one pair of optimal decision variables u_i^2 and v_j^2. This clearly indicates that the existence of more-for-less solution is dependent upon the optimal values of the decision variables u_i^2 and v_j^2. However, as discussed in Sect. 3.9.1 that the optimal values of the decision variables u_i^2 and v_j^2, obtained by the existing methods [2, 6], are not correct. Hence, more-for-less solution obtained by the existing methods [2, 6] may or may not be correct.

3.10 Conclusions

In this chapter, the sufficient condition-II is proposed to check the existence of at least one more-for-less solution of symmetric triangular fuzzy transportation problems with mixed constraints. Also, a method is proposed to find all more-for-less solutions (if exist) of symmetric triangular fuzzy transportation problems with mixed constraints. Furthermore, it is pointed out that a more-for-less solution, obtained by the existing methods [2, 6], may or may not be correct. While, all more-for-less solutions, obtained by the proposed Mehar method-II, will always be correct.

References

1. G.M. Appa, The transportation problem and its variants. J. Operat. Res. Soc. **24**, 79–99 (1973)
2. V. Adlakha, K. Kowalski, B. Lev, Solving transportation problems with mixed constraints. Inter. J. Management Sci. Eng. Manag. **1**, 47–52 (2006)
3. M.E.B. Brigden, A variant of the transportation problem in which the constraints are of mixed type. J. Operat. Res. Soc. **25**, 437–445 (1974)
4. D. Klingman, R. Russell, The transportation problem with mixed constraints. J. Operat. Res. Soc. **25**, 447–455 (1974)
5. H.A. Taha, in *Operations Research: An Introduction* (Pearson Prentice Hall, Upper Saddle River, NJ, 2013)
6. V. Vidhya, P. Uma Maheswari, K. Ganesan, An alternate method for finding more for less solution to fuzzy transportation problem with mixed constraints, Soft Comput. **25**, 11989–11996 (2021)

Chapter 4
Mehar Method-III to Find All More-for-Less Solutions of Symmetric Intuitionistic Fuzzy Transportation Problems with Mixed Constraints

In this chapter, firstly, the arithmetic operations of triangular fuzzy numbers as well as the method for comparing triangular fuzzy numbers, proposed by Vidhya et al. [3], are extended for triangular intuitionistic fuzzy numbers. Then, using the extended arithmetic operations and the extended comparing method, a new method (named as Mehar method-III) is proposed to find all more-for-less solutions of symmetric triangular intuitionistic fuzzy transportation problems with mixed constraints (transportation problems with mixed constraints in which each parameter and each variable is represented by a symmetric triangular intuitionistic fuzzy number). Finally, a symmetric triangular intuitionistic fuzzy transportation problem with mixed constraints is solved to illustrate the proposed Mehar method-III.

4.1 Some Basic Definitions

In this section, some basic definitions are discussed.

Definition 4.1.1 [1] Let X be a universal set. Then, the set $\tilde{A} = \{x, \mu_{\tilde{A}}(x), \upsilon_{\tilde{A}}(x); x \in X\}$ is said to be an intuitionistic fuzzy set defined over the universal set X, where

(i) $\mu_{\tilde{A}} : X \to [0, 1]$ and $\upsilon_{\tilde{A}} : X \to [0, 1]$ are said to be the membership function and non-membership function respectively.
(ii) The values $\mu_{\tilde{A}}(x)$ and $\upsilon_{\tilde{A}}(x)$ are called the degree of membership and degree of non-membership for $x \in \tilde{A}$ respectively.
(iii) The values $\mu_{\tilde{A}}(x)$ and $\upsilon_{\tilde{A}}(x)$ satisfy the condition $0 \leq \mu_{\tilde{A}}(x) + \upsilon_{\tilde{A}}(x) \leq 1$.
(iv) $\tilde{A} = (\mu_{\tilde{A}}, \upsilon_{\tilde{A}})$ is called an intuitionistic fuzzy number.

Definition 4.1.2 [1] An intuitionistic fuzzy number $\tilde{A}$ is said to be a triangular intuitionistic fuzzy number if its membership function and non-membership function

T. K. Bhatia et al., *More-for-Less Solutions in Fuzzy Transportation Problems*, Studies in Fuzziness and Soft Computing 426,
https://doi.org/10.1007/978-3-031-30337-1_4

are defined as

$$\mu_{\tilde{A}}(x) = \begin{cases} \frac{x-a^1}{a^2-a^1}, a^1 < x \le a^2, \\ \frac{a^3-x}{a^3-a^2}, a^2 \le x < a^3, \\ 0, otherwise \end{cases}$$

$$\upsilon_{\tilde{A}}(x) = \begin{cases} \frac{a^2-x}{a^2-A^1}, A^1 < x \le a^2, \\ \frac{x-a^2}{A^3-a^2}, a^2 \le x < A^3, \\ 0, otherwise \end{cases}$$

A triangular intuitionistic fuzzy number $\tilde{A}$ may be represented as $\tilde{A} = \left(a^1, a^2, a^3; A^1, a^2, A^3\right)$ or $\tilde{A} = \left(a^2, a^2 - a^1, a^3 - a^2; a^2, a^2 - A^1, A^3 - a^2\right)$ or $\tilde{A} = \left(a^2, \beta(1-\alpha), \gamma(1-\alpha); a^2, \beta'(1-\alpha), \gamma'(1-\alpha)\right)$ where $\alpha \in (0, 1]$, $\beta = a^2 - a^1$, $\gamma = a^3 - a^2$, $\beta' = a^2 - A^1$ and $\gamma' = A^3 - a^2$.

Definition 4.1.3 [3] A triangular intuitionistic fuzzy number $\tilde{A} = \left(a^2, \beta(1-\alpha), \gamma(1-\alpha); a^2, \beta'(1-\alpha), \gamma'(1-\alpha)\right)$, where $\alpha \in (0, 1]$ is said to be symmetric triangular intuitionistic fuzzy number if $\beta = \gamma$ and $\beta' = \gamma'$. Assuming $\beta = \gamma = \delta$ and $\beta' = \gamma' = \delta'$, a symmetric triangular intuitionistic fuzzy number can be represented as $\tilde{A} = \left(a^2, \delta(1-\alpha), \delta(1-\alpha); a^2, \delta'(1-\alpha), \delta'(1-\alpha)\right)$, where $\alpha \in (0, 1]$.

Definition 4.1.4 [3] A symmetric triangular intuitionistic fuzzy number $\tilde{A} = \left(a^2, \beta(1-\alpha), \gamma(1-\alpha); a^2, \beta'(1-\alpha), \gamma'(1-\alpha)\right)$, where $\alpha \in (0, 1]$ is said to be zero triangular intuitionistic fuzzy number if $a^2 = 0$. Hence, a zero triangular intuitionistic fuzzy number can be represented as $\tilde{A} = \left(0, \delta(1-\alpha), \delta(1-\alpha); 0, \delta'(1-\alpha), \delta'(1-\alpha)\right)$, where δ and δ' are any non-negative real number and $\alpha \in (0, 1]$.

4.2 Extended Arithmetic Operations of Triangular Intuitionistic Fuzzy Numbers

In this section, the existing arithmetic operations of triangular fuzzy numbers [3] is extended for triangular intuitionistic fuzzy numbers.

(i) Let $\tilde{A}_i = \left(a_i^2, \beta_i(1-\alpha), \gamma_i(1-\alpha); a_i^2, \beta_i^{'}(1-\alpha), \gamma_i^{'}(1-\alpha)\right); i = 1, 2, \ldots, k$, where $\alpha \in (0, 1]$, $\beta_i = a_i^2 - a_i^1$, $\gamma_i = a_i^3 - a_i^2$, $\beta_i^{'} = a_i^2 - A_i^1$ and $\gamma_i^{'} = A_i^3 - a_i^2$; $i = 1, 2, \ldots, k$ be triangular intuitionistic fuzzy numbers. Then,

$$\oplus_{i=1}^{k} \tilde{A}_i = \left(\sum_{i=1}^{k} a_i^2, \underset{1 \le i \le k}{\text{maximum}}\{\beta_i(1-\alpha)\}, \underset{1 \le i \le k}{\text{maximum}}\{\gamma_i(1-\alpha)\}; \right.$$

$$\sum_{i=1}^{k} a_i^2, \underset{1\le i\le k}{\text{maximum}}\left\{\beta_i'(1-\alpha)\right\}, \underset{1\le i\le k}{\text{maximum}}\left\{\gamma_i'(1-\alpha)\right\}\Bigg). \quad (4.1)$$

and

$$\otimes_{i=1}^{k}\tilde{A}_i = \Bigg(\prod_{i=1}^{k} a_i^2, \underset{1\le i\le k}{\text{maximum}}\{\beta_i(1-\alpha)\}, \underset{1\le i\le k}{\text{maximum}}\{\gamma_i(1-\alpha)\};$$
$$\prod_{i=1}^{k} a_i^2, \underset{1\le i\le k}{\text{maximum}}\left\{\beta_i'(1-\alpha)\right\}, \underset{1\le i\le k}{\text{maximum}}\left\{\gamma_i'(1-\alpha)\right\}\Bigg). \quad (4.2)$$

(ii) Let $\tilde{A}_1 = \left(a_1^2, \beta_1(1-\alpha), \gamma_1(1-\alpha); a_1^2, \beta_1'(1-\alpha), \gamma_1'(1-\alpha)\right)$ and $\tilde{A}_2 = \left(a_2^2, \beta_2(1-\alpha), \gamma_2(1-\alpha); a_2^2, \beta_2'(1-\alpha), \gamma_2'(1-\alpha)\right); a_2^2 \neq 0$, where $\alpha \in (0, 1]$, $\beta_i = a_i^2 - a_i^1$, $\gamma_i = a_i^3 - a_i^2$, $\beta_i' = a_i^2 - A_i^1$ and $\gamma_i' = A_i^3 - a_i^2; i = 1, 2$ be two triangular intuitionistic fuzzy numbers. Then,

$$\tilde{A}_1 \oslash \tilde{A}_2 = \Bigg(\frac{a_1^2}{a_2^2}, \text{maximum}\{\beta_1(1-\alpha), \beta_2(1-\alpha)\},$$
$$\text{maximum}\{\gamma_1(1-\alpha), \gamma_2(1-\alpha)\};$$
$$\frac{a_1^2}{a_2^2}, \text{maximum}\left\{\beta_1'(1-\alpha), \beta_2'(1-\alpha)\right\},$$
$$\text{maximum}\left\{\gamma_1'(1-\alpha), \gamma_2'(1-\alpha)\right\}\Bigg) \quad (4.3)$$

4.3 Extended Method for Comparing Triangular Intuitionistic Fuzzy Numbers

In this section, the existing method for comparing triangular fuzzy numbers [3] is extended for comparing triangular intuitionistic fuzzy numbers.

Let $\tilde{A}_1 = \left(a_1^2, \beta_1(1-\alpha), \gamma_1(1-\alpha); a_1^2, \beta_1'(1-\alpha), \gamma_1'(1-\alpha)\right)$ and $\tilde{A}_2 = \left(a_2^2, \beta_2(1-\alpha), \gamma_2(1-\alpha); a_2^2, \beta_2'(1-\alpha), \gamma_2'(1-\alpha)\right)$ where $\alpha \in (0, 1]$, $\beta_i = a_i^2 - a_i^1$, $\gamma_i = a_i^3 - a_i^2$, $\beta_i' = a_i^2 - A_i^1$ and $\gamma_i' = A_i^3 - a_i^2; i = 1, 2$ be two triangular intuitionistic fuzzy numbers. Then,

(i) $\tilde{A}_1 <_{\text{Mag}} \tilde{A}_2 \text{if} \text{Mag}\left(\tilde{A}_1\right) < \text{Mag}\left(\tilde{A}_2\right)$,

(ii) $\tilde{A}_1 \le_{\text{Mag}} \tilde{A}_2 \text{if} \text{Mag}\left(\tilde{A}_1\right) \le \text{Mag}\left(\tilde{A}_2\right)$,

(iii) $\tilde{A}_1 =_{\text{Mag}} \tilde{A}_2 \text{if} \text{Mag}\left(\tilde{A}_1\right) = \text{Mag}\left(\tilde{A}_2\right)$.

where,

$$\text{Mag}\left(\tilde{A}_i\right) = \frac{\gamma_i(1-\alpha) + 8a_i^2 - \beta_i(1-\alpha) + \gamma_i^{'}(1-\alpha) - \beta_i^{'}(1-\alpha)}{8}; i = 1, 2. \tag{4.4}$$

4.4 Some Important Results

In this section, it is proved that if $\tilde{A}_i = \left(a_i^2, \delta_i(1-\alpha), \delta_i(1-\alpha); a_i^2, \delta_i'(1-\alpha), \delta_i'(1-\alpha)\right)$; $i = 1, 2, \ldots, k,$, where $\alpha \in (0, 1]$, $\delta_i = a_i^2 - a_i^1 = a_i^3 - a_i^2$, $\delta_i^{'} = a_i^2 - A_i^1 = A_i^3 - a_i^2$ are symmetric triangular intuitionistic fuzzy numbers. Then,

$$\text{Mag}\left(\tilde{A}_i\right) = a_i^2 \tag{4.5}$$

$$\text{Mag}\left(\oplus_{i=1}^k \tilde{A}_i\right) = \sum_{i=1}^k \text{Mag}\left(\tilde{A}_i\right) \tag{4.6}$$

$$\text{Mag}\left(\otimes_{i=1}^k \tilde{A}_i\right) = \prod_{i=1}^k \text{Mag}\left(\tilde{A}_i\right) \tag{4.7}$$

Also, it is proved that if $\tilde{A}_1 = \left(a_1^2, \delta_1(1-\alpha), \delta_1(1-\alpha); a_1^2, \delta_1^{'}(1-\alpha), \delta_1^{'}(1-\alpha)\right)$ and $\tilde{A}_2 = \left(a_2^2, \delta_2(1-\alpha), \delta_2(1-\alpha); a_2^2, \delta_2^{'}(1-\alpha), \delta_2^{'}(1-\alpha)\right)$; $a_2^2 \neq 0$, where $\alpha \in (0, 1]$, $\delta_i = a_i^2 - a_i^1 = a_i^3 - a_i^2, \delta_i^{'} = a_i^2 - A_i^1 = A_i^3 - a_i^2; i = 1, 2$ are two triangular intuitionistic fuzzy numbers. Then,

$$\text{Mag}\left(\tilde{A}_1 \oslash \tilde{A}_2\right) = \frac{\text{Mag}\left(\tilde{A}_1\right)}{\text{Mag}\left(\tilde{A}_2\right)} \tag{4.8}$$

4.4.1 Proof of the First Result

Using the expression (4.4),

$$\text{Mag}\left(\tilde{A}_i\right) = \frac{\delta_i(1-\alpha) + 8a_i^2 - \delta_i(1-\alpha) + \delta_i^{'}(1-\alpha) - \delta_i^{'}(1-\alpha)}{8} = a_i^2 \tag{4.9}$$

4.4.2 Proof of the Second Result

Using the expression (4.1),

$$\oplus_{i=1}^{k}\tilde{A}_i = \left(\sum_{i=1}^{k} a_i^2, \max_{1\le i\le k}\text{imum}\{\delta_i(1-\alpha)\}, \max_{1\le i\le k}\text{imum}\{\delta_i(1-\alpha)\};\right.$$
$$\left.\sum_{i=1}^{k} a_i^2, \max_{1\le i\le k}\text{imum}\left\{\delta_i^{'}(1-\alpha)\right\}, \max_{1\le i\le k}\text{imum}\left\{\delta_i^{'}(1-\alpha)\right\}\right).$$

Using the expression (4.4),

$$\text{Mag}\left(\oplus_{i=1}^{n}\tilde{A}_i\right) = \frac{1}{8}\left(\underset{1\le i\le k}{\text{maximum}}\{\delta_i(1-\alpha)\} + 8\sum_{i=1}^{k} a_i^2 - \underset{1\le i\le k}{\text{maximum}}\{\delta_i(1-\alpha)\}\right.$$
$$\left. + \underset{1\le i\le k}{\text{maximum}}\left\{\delta_i^{'}(1-\alpha)\right\} - \underset{1\le i\le k}{\text{maximum}}\left\{\delta_i^{'}(1-\alpha)\right\}\right);$$
$$i = 1, 2, \ldots, k.$$

$$= \sum_{i=1}^{k} a_i^2 \tag{4.10}$$

Using the expression (4.9),

$$\text{Mag}\left(\oplus_{i=1}^{k}\tilde{A}_i\right) = \sum_{i=1}^{k}\text{Mag}\left(\tilde{A}_i\right).$$

4.4.3 Proof of the Third Result

Using the expression (4.2),

$$\otimes_{i=1}^{k}\tilde{A}_i = \left(\prod_{i=1}^{k} a_i^2, \max_{1\le i\le k}\text{imum}\{\delta_i(1-\alpha)\}, \max_{1\le i\le k}\text{imum}\{\delta_i(1-\alpha)\};\right.$$
$$\left.\prod_{i=1}^{k} a_i^2, \max_{1\le i\le k}\text{imum}\left\{\delta_i^{'}(1-\alpha)\right\}, \max_{1\le i\le k}\text{imum}\left\{\delta_i^{'}(1-\alpha)\right\}\right).$$

Using the expression (4.4),

$$\begin{aligned}\text{Mag}\left(\otimes_{i=1}^{k}\tilde{A}_i\right) =& \frac{1}{8}\left(\underset{1\le i\le k}{\max\text{imum}}\{\delta_i(1-\alpha)\} + 8\prod_{i=1}^{k} a_i^2 - \underset{1\le i\le k}{\max\text{imum}}\{\delta_i(1-\alpha)\}\right.\\ & + \underset{1\le i\le k}{\max\text{imum}}\left\{\delta_i^{'}(1-\alpha)\right\}\\ & \left. - \underset{1\le i\le k}{\max\text{imum}}\left\{\delta_i^{'}(1-\alpha)\right\}\right); i = 1, 2, \ldots, k.\\ & = \prod_{i=1}^{k} a_i^2 \end{aligned} \tag{4.11}$$

Using the expression (4.9),

$$\text{Mag}\left(\otimes_{i=1}^{k}\tilde{A}_i\right) = \prod_{i=1}^{k}\text{Mag}\left(\tilde{A}_i\right).$$

4.4.4 *Proof of the Fourth Result*

Using the expression (4.3),

$$\begin{aligned}\tilde{A}_1 \oslash \tilde{A}_2 =& \left(\frac{a_1^2}{a_2^2}, \text{maximum}\{\delta_1(1-\alpha), \delta_2(1-\alpha)\},\right.\\ & \text{maximum}\{\delta_1(1-\alpha), \delta_2(1-\alpha)\};\\ & \frac{a_1^2}{a_2^2}, \text{maximum}\left\{\delta_1^{'}(1-\alpha), \delta_2^{'}(1-\alpha)\right\},\\ & \left.\text{maximum}\left\{\delta_1^{'}(1-\alpha), \delta_2^{'}(1-\alpha)\right\}\right).\end{aligned}$$

Using the expression (4.4),

$$\begin{aligned}\text{Mag}\left(\tilde{A}_1 \oslash \tilde{A}_2\right) =& \frac{1}{8}\left(\text{maximum}\left\{\delta_1(1-\alpha), \delta_2^{'}(1-\alpha)\right\} + 8\left(\frac{a_1^2}{a_2^2}\right)\right.\\ & - \text{maximum}\left\{\delta_1(1-\alpha), \delta_2^{'}(1-\alpha)\right\}\\ & + \text{maximum}\left\{\delta_1(1-\alpha), \delta_2^{'}(1-\alpha)\right\}\\ & \left. - \text{maximum}\left\{\delta_1(1-\alpha), \delta_2^{'}(1-\alpha)\right\}\right) = \frac{a_1^2}{a_2^2}\end{aligned} \tag{4.12}$$

Using the expression (4.9),

$$\text{Mag}\left(\tilde{A}_1 \oslash \tilde{A}_2\right) = \frac{\text{Mag}\left(\tilde{A}_1\right)}{\text{Mag}\left(\tilde{A}_2\right)}.$$

4.5 Tabular Representation of Symmetric Triangular Intuitionistic Fuzzy Transportation Problems with Mixed Constraints

A symmetric triangular intuitionistic fuzzy transportation problem with mixed constraints can be represented into tabular form as shown by Table 4.1.

where,

(i) The symmetric triangular intuitionistic fuzzy number $\tilde{a}_i = \left(a_i^2, \delta_{i1}(1-\alpha), \delta_{i1}(1-\alpha); a_i^2, \delta_{i2}(1-\alpha), \delta_{i2}(1-\alpha)\right)$ represents the supply of product at the ith source.

(ii) The symmetric triangular intuitionistic fuzzy number $\tilde{b}_j = \left(b_j^2, \delta_{j3}(1-\alpha), \delta_{j3}(1-\alpha); b_j^2, \delta_{j4}(1-\alpha), \delta_{j4}(1-\alpha)\right)$ represents the demand of product at the jth destination.

(iii) The symmetric triangular intuitionistic fuzzy number $\tilde{c}_{ij} = \left(c_{ij}^2, \delta_{ij1}(1-\alpha), \delta_{ij1}(1-\alpha); c_{ij}^2, \delta_{ij2}(1-\alpha), \delta_{ij2}(1-\alpha)\right)$ represents the cost for transporting one unit quantity of product from the ith source to the jth destination.

Table 4.1 Tabular representation of a symmetric triangular intuitionistic fuzzy transportation problem with mixed constraints

	D_1	D_2	…	D_n	Availability
S_1	$\tilde{c}_{11}$	$\tilde{c}_{12}$	…	$\tilde{c}_{1n}$	$=_{\text{Mag}}$ *or* $\leq_{\text{Mag}}$ *or* $\geq_{\text{Mag}}$ $\tilde{a}_1$
S_2	$\tilde{c}_{21}$	$\tilde{c}_{22}$	…	$\tilde{c}_{2n}$	$=_{\text{Mag}}$ *or* $\leq_{\text{Mag}}$ *or* $\geq_{\text{Mag}}$ $\tilde{a}_2$
⋮	⋮	⋮	⋮	⋮	⋮
S_m	$\tilde{c}_{m1}$	$\tilde{c}_{m2}$	…	$\tilde{c}_{mn}$	$=_{\text{Mag}}$ *or* $\leq_{\text{Mag}}$ *or* $\geq_{\text{Mag}}$ $\tilde{a}_m$
Demand	$=_{\text{Mag}}$ *or* $\leq_{\text{Mag}}$ *or* $\geq_{\text{Mag}}$ $\tilde{b}_1$	$=_{\text{Mag}}$ *or* $\leq_{\text{Mag}}$ *or* $\geq_{\text{Mag}}$ $\tilde{b}_2$	…	$=_{\text{Mag}}$ *or* $\leq_{\text{Mag}}$ *or* $\geq_{\text{Mag}}$ $\tilde{b}_n$	

4.6 Intuitionistic Fuzzy Linear Programming Problems Corresponding to Symmetric Triangular Intuitionistic Fuzzy Transportation Problems with Mixed Constraints

If it is assumed that the symmetric triangular intuitionistic fuzzy number $\tilde{x}_{ij} = \left(x_{ij}^2, \delta_{ij3}(1-\alpha), \delta_{ij3}(1-\alpha); x_{ij}^2, \delta_{ij4}(1-\alpha), \delta_{ij4}(1-\alpha)\right)$ represents the fuzzy quantity of the product to be supplied from the ith source to the jth destination, the symmetric triangular intuitionistic fuzzy number $\tilde{Z}$ represents the total intuitionistic fuzzy transportation cost and $\tilde{0}$ represents the zero triangular intuitionistic fuzzy number. Then, the symmetric triangular intuitionistic fuzzy transportation problem with mixed constraints, represented by Table 4.1, can be formulated into its equivalent intuitionistic fuzzy linear programming problem ($P4.1$) or into its equivalent crisp linear programming problem ($P4.2$) (dual problem of the intuitionistic fuzzy linear programming problem ($P4.1$).

Problem ($P4.1$)

$$\text{Minimize}\left(\tilde{Z} =_{\text{Mag}} \oplus_{i=1}^{m} \oplus_{j=1}^{n} \left(\tilde{c}_{ij} \otimes \tilde{x}_{ij}\right)\right)$$

Subject to

$$\oplus_{j=1}^{n} \tilde{x}_{ij} =_{\text{Mag}} or \leq_{\text{Mag}} or \geq_{\text{Mag}} \tilde{a}_i;\ i = 1, 2, \ldots, m;\ \ (i\text{th source constraint})$$
$$\oplus_{i=1}^{m} \tilde{x}_{ij} =_{\text{Mag}} or \leq_{\text{Mag}} or \geq_{\text{Mag}} \tilde{b}_j;\ j = 1, 2, \ldots, n;\ \ (j\text{th destination constraint})$$
$$\tilde{x}_{ij} \geq_{\text{Mag}} \tilde{0};\ i = 1, 2, \ldots, m;\ j = 1, 2, \ldots, n.$$

Problem ($P4.2$)

$$\text{Maximize}\left(Z^2 = \sum_{i=1}^{m} a_i^2 u_i^2 + \sum_{j=1}^{n} b_j^2 v_j^2\right)$$

Subject to

$$u_i^2 + v_j^2 \leq c_{ij}^2;\ i = 1, 2, \ldots, m;\ j = 1, 2, \ldots, n;$$
$$u_i^2 * 0,\ v_j^2 \circ 0;\ i = 1, 2, \ldots, m;\ j = 1, 2, \ldots, n;$$

where,

(i) $*$ will be $\geq$ if the sign of the ith source constraint of the intuitionistic fuzzy linear programming problem ($P4.1$) is $\geq$.
(ii) $*$ will be $\leq$ if the sign of the ith source constraint of the intuitionistic fuzzy linear programming problem ($P4.1$) is $\leq$.
(iii) If the sign of the ith source constraint of the intuitionistic fuzzy linear programming problem ($P4.1$) is $=$. Then, u_i will be an unrestricted variable.

(iv) $\circ$ will be $\geq$ if the sign of the jth destination constraint of the intuitionistic fuzzy linear programming problem ($P4.1$) is $\geq$.
(v) $\circ$ will be $\leq$ if the sign of the jth destination constraint of the intuitionistic fuzzy linear programming problem ($P4.1$) is $\leq$.
(vi) If the sign of the jth destination constraint of the intuitionistic fuzzy linear programming problem ($P4.1$) $=$. Then, v_j will be an unrestricted variable.

The dual problem ($P4.2$) is obtained as follows.

Step 1: Using the comparing method, proposed in Sect. 4.3, the intuitionistic fuzzy linear programming problem ($P4.1$) can be transformed into its equivalent crisp linear programming problem ($P4.3$).

Problem (*P*4.3)

$$\text{Minimize}\left(\text{Mag}\left(\tilde{Z}\right) = \text{Mag}\left(\oplus_{i=1}^{m} \oplus_{j=1}^{n} \left(\tilde{c}_{ij} \otimes \tilde{x}_{ij}\right)\right)\right)$$

Subject to

$$\text{Mag}\left(\oplus_{j=1}^{n}\tilde{x}_{ij}\right) = or \geq or \leq \text{Mag}(\tilde{a}_i);\ i = 1, 2, \ldots, m;$$
$$\text{Mag}\left(\oplus_{i=1}^{m}\tilde{x}_{ij}\right) = or \geq or \leq \text{Mag}\left(\tilde{b}_j\right);\ j = 1, 2, \ldots, n;$$
$$\text{Mag}\left(\tilde{x}_{ij}\right) \geq \text{Mag}\left(\tilde{0}\right);\ i = 1, 2, \ldots, m;\ j = 1, 2, \ldots, n.$$

Step 2: Using the expression (4.4), the crisp linear programming problem ($P4.3$) can be transformed into its equivalent crisp linear programming problem ($P4.4$).

Problem (*P*4.4)

$$\text{Minimize}\left(\text{Mag}\left(\tilde{Z}\right) = \left(\sum_{i=1}^{m}\sum_{j=1}^{n}\text{Mag}\left(\tilde{c}_{ij} \otimes \tilde{x}_{ij}\right)\right)\right)$$

Subject to

$$\sum_{j=1}^{n}\text{Mag}\left(\tilde{x}_{ij}\right) = or \geq or \leq \text{Mag}(\tilde{a}_i);\ i = 1, 2, \ldots, m;$$
$$\sum_{i=1}^{m}\text{Mag}\left(\tilde{x}_{ij}\right) = or \geq or \leq \text{Mag}\left(\tilde{b}_j\right);\ j = 1, 2, \ldots, n$$
$$\text{Mag}\left(\tilde{x}_{ij}\right) \geq \text{Mag}\left(\tilde{0}\right);\ i = 1, 2, \ldots, m;\ j = 1, 2, \ldots, n.$$

Step 3: Using the expression (4.5), the crisp linear programming problem ($P4.4$) can be transformed into its equivalent crisp linear programming problem ($P4.5$).

Problem (*P*4.5)

$$\text{Minimize}\left(\text{Mag}\left(\tilde{Z}\right)=\left(\sum_{i=1}^{m}\sum_{j=1}^{n}\text{Mag}(\tilde{c}_{ij})\text{Mag}(\tilde{x}_{ij})\right)\right)$$

Subject to

Constraints of the problem (*P*4.4).

Step 4: Using the expression (4.2), transform the crisp linear programming problem (*P*4.5) into its equivalent crisp linear programming problem (*P*4.6).

Problem (*P*4.6)

$$\text{Minimize}\left(Z=\sum_{i=1}^{m}\sum_{j=1}^{n}c_{ij}^{2}x_{ij}^{2}\right)$$

Subject to

$$\sum_{j=1}^{n}x_{ij}^{2} = or \geq or \leq a_i^2;\ i=1,2,\ldots,m;$$
$$\sum_{i=1}^{m}x_{ij}^{2} = or \geq or \leq b_j^2;\ j=1,2,\ldots,n;$$
$$x_{ij}^{2}\geq 0;\ i=1,2,\ldots,m;\ j=1,2,\ldots,n.$$

Step 5: Using the duality theory [2], the crisp linear programming problem (*P*4.2) represents the dual of the crisp linear programming problem (*P*4.6). Hence, the crisp linear programming problem (*P*4.2) represents the dual of the intuitionistic fuzzy linear programming problem (*P*4.1).

4.7 Crisp Transportation Problem Equivalent to Symmetric Intuitionistic Fuzzy Transportation Problem

The following clearly indicates that the symmetric triangular intuitionistic fuzzy transportation problem with mixed constraints, represented by Table 4.1, is equivalent to crisp transportation problem with mixed constraints represented by Table 4.2.

It is obvious from Step 1 to Step 4 of Sect. 4.6 that the intuitionistic fuzzy linear programming problem (*P*4.1) corresponding to the symmetric triangular intuitionistic fuzzy transportation problem with mixed constraints, represented by Table 4.1, is equivalent to the crisp linear programming problem (*P*4.6). Also, it is obvious from Sect. 3.4 of Chap. 3 that the crisp linear programming problem (*P*4.6) is equivalent to crisp transportation problem with mixed constraints represented by Table 4.2.

Table 4.2 Crisp transportation problem with mixed constraints

	D_1	D_2	...	D_n	Availability
S_1	c_{11}^2	c_{12}^2	...	c_{1n}^2	$= or \leq or \geq a_1^2$
S_2	c_{21}^2	c_{22}^2	...	c_{2n}^2	$= or \leq or \geq a_2^2$
⋮	⋮	⋮	⋮	⋮	⋮
S_m	c_{m1}^2	c_{m2}^2	...	c_{mn}^2	$= or \leq or \geq a_m^2$
Demand	$= or \leq or \geq b_1^2$	$= or \leq or \geq b_2^2$	...	$= or \leq or \geq b_n^2$	

4.8 Proposed Sufficient Condition-III for the Existence of at Least One More-for-Less Solution

It is obvious from Sect. 4.7 that the symmetric triangular intuitionistic fuzzy transportation problem with mixed constraints, represented by Table 4.1, is equivalent to the crisp transportation problem with mixed constraints represented by Table 4.2. Hence, to propose a sufficient condition-III for the existence of at least one more-less-solution for the symmetric triangular intuitionistic fuzzy transportation problem with mixed constraints, represented by Table 4.1, is equivalent to propose a sufficient condition-III for the existence of at least one more-less-solution for the crisp transportation problem with mixed constraints represented by Table 4.2. Hence, the following sufficient condition-III to check the existence of at least one more-for-less solution of crisp transportation problem with mixed constraints, proposed in Sect. 3.7 of Chap. 3, can be used to check the existence of at least one more-less-solution of symmetric triangular intuitionistic fuzzy transportation problem with mixed constraints represented by Table 4.1.

"If in an optimal solution of the crisp linear programming problem ($P4.2$), there exist variables u_p^2 and v_q^2 such that $u_p^2 + v_q^2 < 0$. Then, at least one more-for-less solution of crisp transportation problem with mixed constraints, represented by Table 4.2, will exist. Hence, at least one more-for-less solution of symmetric triangular fuzzy transportation problem with mixed constraints, represented by Table 4.1, will exist."

4.9 Proposed Mehar Method-III

In this section, a new method (named as Mehar method-III) is proposed to check the existence of at least one more-for-less solution as well as to find all more-for-less solutions (if exists) of symmetric triangular intuitionistic fuzzy transportation problems with mixed constraints.

The steps of the proposed Mehar method-III are as follows.

Step 1: Use **Step 1.1** to **Step 1.4** to check that for the symmetric triangular intuitionistic fuzzy transportation problem with mixed constraints, represented by Table 4.1, at least one more-for-less solution will exist or not.

Step 1.1: Transform the symmetric triangular intuitionistic fuzzy transportation problem with mixed constraints, represented by Table 4.1, into its equivalent crisp transportation problem with mixed constraints, represented by Table 4.2.
Step 1.2: Find an optimal solution $\left\{u_i^2; i = 1, 2, \ldots, m, v_j^2; j = 1, 2, \ldots, n\right\}$ of the problem ($P4.2$) by substituting the values of c_{ij}^2, a_i^2 and b_j^2 from Table 4.2.
Step 1.3: Using the optimal values of $u_i^2; i = 1, 2, \ldots, m, v_j^2; j = 1, 2, \ldots, n$, obtained in **Step 1.2**, construct a shadow price matrix $\left[A_{ij}\right]_{m \times n}$, where $A_{ij} = u_i^2 + v_j^2$.
Step 1.4: Check that all the elements A_{ij} of the shadow price matrix, obtained in **Step 1.3**, are non-negative or not.

Case (i): If all the elements of the shadow price matrix, obtained in **Step 1.3**, are non-negative. Then, there does not exist a more-for-less solution.
Case (ii): If at least one element of the shadow price matrix, obtained in **Step 1.3**, is negative. Then, at least one more-for-less solution will exist.

Step 2: If the element corresponding to the pth row and the qth column of the shadow price matrix, obtained in **Step 1.3**, is negative i.e., if $u_p^2 + v_q^2 < 0$. Then, a more-for-less solution can be obtained by increasing the availability of the pth source and the demand of the qth destination with the help of **Step 2.1** to **Step 2.8** of the post-optimality analysis [2].

Step 2.1: Construct the matrix

$$R_{pq} = \begin{bmatrix} a_1 \\ a_2 \\ \vdots \\ a_p + \theta \\ \vdots \\ a_m \\ b_1 \\ b_2 \\ \vdots \\ b_q + \theta \\ \vdots \\ b_n \end{bmatrix}$$

where,

(i) θ is a positive real number.
(ii) a_i is the availability of the product at the ith source.
(iii) b_j is the demand of the product at the jth destination.

Step 2.2: Identify the basic variables in an optimal solution of the crisp linear programming problem ($P4.7$).

Problem ($P4.7$)

$$\text{Minimize}\left(Z^2 = \sum_{i=1}^{m}\sum_{j=1}^{n} c_{ij}^2 x_{ij}^2\right)$$

Subject to

$$\sum_{j=1}^{n} x_{ij}^2 + P_i s_i^2 = a_i^2;\ i = 1, 2, \ldots, m;$$
$$\sum_{i=1}^{m} x_{ij}^2 + Q_j s_j^2 = b_j^2;\ j = 1, 2, \ldots, n;$$
$$x_{ij}^2, s_i^2, s_j^2 \geq 0;\ i = 1, 2, \ldots, m;\ j = 1, 2, \ldots, n,$$

where,

(i) $P_i = 1$ if the sign of the ith source constraint of the crisp linear programming problem ($P4.6$) is $\leq$.
(ii) $P_i = -1$ if the sign of the ith source constraint of the crisp linear programming problem ($P4.6$) is $\geq$.
(iii) $P_i = 0$ if the sign of the ith source constraint of the crisp linear programming problem ($P4.6$) is $=$.
(iv) $Q_j = 1$ if the sign of the jth destination constraint of the crisp linear programming problem ($P4.6$) is $\leq$.
(v) $Q_j = -1$ if the sign of the jth destination constraint of the crisp linear programming problem ($P4.6$) is $\geq$.
(vi) $Q_j = 0$ if the sign of the jth destination constraint of the crisp linear programming problem ($P4.6$) is $=$.

Step 2.3: Construct the basis matrix B with the help of the coefficients of the identified optimal basic variables in the constraints of the crisp linear programming problem ($P4.7$).
Step 2.4: Find the multiplicative inverse of the basis matrix B i.e., find B^{-1}.
Step 2.5: Find the matrix $B^{-1}R_{pq}$.
Step 2.6: Find the range of θ(where, $\theta > 0$) for which all the elements of the matrix $B^{-1}R_{pq}$, obtained in **Step 2.5**, are greater than or equal to zero.

Step 2.7: For any value of θ lying in its range, obtained in **Step 2.6**, the ith element (say, τ_i) of the matrix $B^{-1}R_{pq}$ represents the value of the ith basic variable in a more-for-less solution of crisp transportation problem with mixed constraints represented by Table 4.2.
Step 2.8: For any value of θ lying in its range, obtained in **Step 2.6**, the symmetric triangular intuitionistic fuzzy number $(\tau_i, \delta_{i1}(1-\alpha), \delta_{i1}(1-\alpha); \tau_i, \delta_{i2}(1-\alpha), \delta_{i2}(1-\alpha))$, where δ_{i1} and δ_{i2} are any non-negative real numbers and $\alpha \in (0, 1]$, represents the value of the ith basic intuitionistic fuzzy variable in a more-for-less solution of the symmetric triangular intuitionistic fuzzy transportation problem with mixed constraints, represented by Table 4.1.

4.10 Illustrative Example

In this section, to illustrate the proposed Mehar method-III, all more-for-less solutions the symmetric triangular intuitionistic fuzzy transportation problem with mixed constraints, represented by Table 4.3, are obtained by the proposed Mehar method-III.

where,

$$\text{(i)}\quad \begin{aligned}
\tilde{c}_{11} &= (3, 1(1-\alpha), 1(1-\alpha); 3, 2(1-\alpha), 2(1-\alpha)),\\
\tilde{c}_{12} &= (5, 2(1-\alpha), 2(1-\alpha); 5, 3(1-\alpha), 3(1-\alpha)),\\
\tilde{c}_{13} &= (5, 2(1-\alpha), 2(1-\alpha); 5, 4(1-\alpha), 4(1-\alpha)),\\
\tilde{c}_{21} &= (5, 1(1-\alpha), 1(1-\alpha); 5, 2(1-\alpha), 2(1-\alpha)),\\
\tilde{c}_{22} &= (9, 3(1-\alpha), 3(1-\alpha); 9, 5(1-\alpha), 5(1-\alpha)),\\
\tilde{c}_{23} &= (4, 2(1-\alpha), 2(1-\alpha); 4, 3(1-\alpha), 3(1-\alpha)),\\
\tilde{c}_{31} &= (4, 1(1-\alpha), 1(1-\alpha); 4, 4(1-\alpha), 4(1-\alpha)),\\
\tilde{c}_{32} &= (2, 1(1-\alpha), 1(1-\alpha); 2, 2(1-\alpha), 2(1-\alpha)),\\
\tilde{c}_{33} &= (4, 1(1-\alpha), 1(1-\alpha); 4, 2(1-\alpha), 2(1-\alpha)).
\end{aligned}$$

$$\text{(ii)}\quad \begin{aligned}
\tilde{a}_1 &= (3, 1(1-\alpha), 1(1-\alpha); 3, 2(1-\alpha), 2(1-\alpha)),\\
\tilde{a}_2 &= (15, 5(1-\alpha), 5(1-\alpha); 15, 10(1-\alpha), 10(1-\alpha)),\\
\tilde{a}_3 &= (10, 4(1-\alpha), 4(1-\alpha); 10, 8(1-\alpha), 8(1-\alpha)).
\end{aligned}$$

Table 4.3 Symmetric triangular intuitionistic fuzzy transportation problem with mixed constraints

	D_1	D_2	D_3	Availability
S_1	$\tilde{c}_{11}$	$\tilde{c}_{12}$	$\tilde{c}_{13}$	$\leq_{\text{Mag}} \tilde{a}_1$
S_2	$\tilde{c}_{21}$	$\tilde{c}_{22}$	$\tilde{c}_{23}$	$=_{\text{Mag}} \tilde{a}_2$
S_3	$\tilde{c}_{31}$	$\tilde{c}_{32}$	$\tilde{c}_{33}$	$\leq_{\text{Mag}} \tilde{a}_3$
Demand	$=_{\text{Mag}} \tilde{b}_1$	$=_{\text{Mag}} \tilde{b}_2$	$=_{\text{Mag}} \tilde{b}_3$	

$$\text{(iii)}\quad \begin{aligned}\tilde{b}_1 &=(6, 2(1-\alpha), 2(1-\alpha); 6, 5(1-\alpha), 5(1-\alpha)),\\ \tilde{b}_2 &=(10, 2(1-\alpha), 2(1-\alpha); 10, 5(1-\alpha), 5(1-\alpha)),\\ \tilde{b}_3 &=(5, 1(1-\alpha), 1(1-\alpha); 5, 3(1-\alpha), 3(1-\alpha)).\end{aligned}$$

Using the proposed Mehar method-III, all more-for-less solutions (if exists) of the symmetric triangular intuitionistic fuzzy transportation problem with mixed constraints, represented by Table 4.3, can be obtained as follows.

Step 1: According to **Step 1** of the proposed Mehar method-III, there is a need to use **Step 1.1** to **Step 1.4** to check that for the symmetric triangular intuitionistic fuzzy transportation problem with mixed constraints, represented by Table 4.3, at least one more-for-less solution will exist or not.

Step 1.1: According to **Step 1.1** of the proposed Mehar method-III, the symmetric triangular intuitionistic fuzzy transportation problem with mixed constraints, represented by Table 4.3, can be transformed into its equivalent crisp transportation problem with mixed constraints represented by Table 4.4.

Step 1.2: According to **Step 1.2** of the proposed Mehar method-III, there is a need to find an optimal solution of the crisp linear programming problem (*P*4.8).

Problem (*P*4.8)

$$\text{Maximize}\left(3u_1^2 + 15u_2^2 + 10u_3^2 + 6v_1^2 + 10v_2^2 + 5v_3^2\right)$$

Subject to

$$\begin{aligned} &u_1^2 + v_1^2 \leq 3,\\ &u_1^2 + v_2^2 \leq 5,\\ &u_1^2 + v_3^2 \leq 5,\\ &u_2^2 + v_1^2 \leq 5,\\ &u_2^2 + v_2^2 \leq 9,\\ &u_2^2 + v_3^2 \leq 4,\\ &u_3^2 + v_1^2 \leq 4,\\ &u_3^2 + v_2^2 \leq 2,\end{aligned}$$

Table 4.4 Transformed crisp transportation problem with mixed constraints

	D_1	D_2	D_3	Availability
S_1	3	5	5	≤ 3
S_2	5	9	4	$= 15$
S_3	4	2	4	≤ 10
Demand	$= 6$	$= 10$	$= 5$	

$$u_3^2 + v_3^2 \leq 4,$$
$$u_1^2, u_3^2 \leq 0,$$
$$u_2^2, v_1^2, v_2^2, v_3^2$$

are unrestricted in sign.

It can be easily verified that on solving the problem ($P4.8$), the obtained optimal solution is $u_1^2 = 0$, $u_2^2 = 7$, $u_3^2 = 0$, $v_1^2 = -2$, $v_2^2 = 2$, $v_3^2 = -3$.

Step 1.3: According to **Step 1.3** of the proposed Mehar method-III, the following shadow price matrix is obtained.

$$\begin{bmatrix} u_1^2 + v_1^2 = -2 & u_1^2 + v_2^2 = 2 & u_1^2 + v_3^2 = -3 \\ u_2^2 + v_1^2 = 5 & u_2^2 + v_2^2 = 9 & u_2^2 + v_3^2 = 4 \\ u_3^2 + v_1^2 = -2 & u_3^2 + v_2^2 = 2 & u_3^2 + v_3^2 = -3 \end{bmatrix}$$

Step 1.4: Since, some elements of the shadow price matrix, obtained in **Step 1.3**, are negative. So, according to Case (ii) of **Step 1.4** of the proposed Mehar method-III, there will exist at least one more-for-less solution of the symmetric triangular intuitionistic fuzzy transportation problem with mixed constraints, represented by Table 4.3.

Step 2: According to Step 2 of the proposed Mehar method-III, all more-for-less solutions of the crisp transportation problem with mixed constraints, represented by Table 4.4, can be obtained as follows.

Step 2.1: It is obvious from **Step 1.3** that $u_1^2+v_1^2 < 0$, $u_1^2+v_3^2 < 0$, $u_3^2+v_1^2 < 0$ and $u_3^2 + v_3^2 < 0$. Therefore, according to **Step 2.1** of the proposed Mehar method-III,

(i) $R_{11} = \begin{bmatrix} 3+\theta \\ 15 \\ 10 \\ 6+\theta \\ 10 \\ 5 \end{bmatrix}$

(ii) $R_{13} = \begin{bmatrix} 3+\theta \\ 15 \\ 10 \\ 6 \\ 10 \\ 5+\theta \end{bmatrix}$

(iii) $R_{31} = \begin{bmatrix} 3 \\ 15 \\ 10+\theta \\ 6+\theta \\ 10 \\ 5 \end{bmatrix}$

(iv) $R_{33} = \begin{bmatrix} 3 \\ 15 \\ 10+\theta \\ 6 \\ 10 \\ 5+\theta \end{bmatrix}$

Step 2.2: According to **Step 2.2** of the proposed Mehar method-III, there is a need to identify the basic variables in an optimal solution of the crisp linear programming problem ($P4.9$).

Problem ($P4.9$)

$$\text{Minimize}\left(Z = 3x_{11}^2 + 5x_{12}^2 + 5x_{13}^2 + 5x_{21}^2 + 9x_{22}^2 + 4x_{23}^2 + 4x_{31}^2 + 2x_{32}^2 + 4x_{33}^2\right)$$

Subject to

$$\begin{aligned}
&x_{11}^2 + x_{12}^2 + x_{13}^2 + s_1 = 3,\\
&x_{21}^2 + x_{22}^2 + x_{23}^2 = 15,\\
&x_{31}^2 + x_{32}^2 + x_{33}^2 + s_2 = 10,\\
&x_{11}^2 + x_{21}^2 + x_{31}^2 = 6,\\
&x_{12}^2 + x_{22}^2 + x_{32}^2 = 10,\\
&x_{13}^2 + x_{23}^2 + x_{33}^2 = 5,\\
&x_{ij}^2 \geq 0;\ i = 1, 2, 3;\ j = 1, 2, 3;\ s_k \geq 0;\ k = 1, 2.
\end{aligned}$$

It can be easily verified that the following optimal solution is obtained on solving the crisp linear programming problem ($P4.9$).

$$\begin{aligned}
&x_{11}^2 = 0,\ x_{12}^2 = 0,\ x_{13}^2 = 0,\ x_{21}^2 = 6,\ x_{22}^2 = 4,\\
&x_{23}^2 = 5,\ x_{31}^2 = 0,\ x_{32}^2 = 6,\ x_{33}^2 = 0,\ s_1^2 = 3,\ s_2^2 = 4.
\end{aligned}$$

Also, it can be easily verified that the basic variables in this obtained optimal solution are x_{21}^2, x_{22}^2, x_{23}^2, x_{32}^2, s_1^2 and s_2^2.

Step 2.3: According to **Step 2.3** of the proposed Mehar method-III, there is a need to construct the basis matrix B with the help of the coefficients of optimal basic variables $x_{21}^2, x_{22}^2, x_{23}^2, x_{32}^2, s_1^2, s_2^2$ in the constraints of the crisp linear programming problem ($P4.9$).

$$B = \begin{bmatrix} 0 & 0 & 0 & 0 & 1 & 0 \\ 1 & 1 & 1 & 0 & 0 & 0 \\ 0 & 0 & 0 & 1 & 0 & 1 \\ 1 & 0 & 0 & 0 & 0 & 0 \\ 0 & 1 & 0 & 1 & 0 & 0 \\ 0 & 0 & 1 & 0 & 0 & 0 \end{bmatrix}$$

Step 2.4: According to **Step 2.4** of the proposed Mehar method-III, there is a need to find the multiplicative inverse of the basis matrix B.

$$B^{-1} = \begin{bmatrix} 0 & 0 & 0 & 1 & 0 & 0 \\ 0 & 1 & 0 & -1 & 0 & -1 \\ 0 & 0 & 0 & 0 & 0 & 1 \\ 0 & -1 & 0 & 1 & 1 & 1 \\ 1 & 0 & 0 & 0 & 0 & 0 \\ 0 & 1 & 1 & -1 & -1 & -1 \end{bmatrix}$$

Step 2.5: According to **Step 2.5** of the proposed Mehar method-III, there is a need to find $B^{-1}R_{11}$, $B^{-1}R_{13}$, $B^{-1}R_{31}$ and $B^{-1}R_{33}$.

$$\text{(i)} \quad B^{-1}R_{11} = \begin{bmatrix} 0 & 0 & 0 & 1 & 0 & 0 \\ 0 & 1 & 0 & -1 & 0 & -1 \\ 0 & 0 & 0 & 0 & 0 & 1 \\ 0 & -1 & 0 & 1 & 1 & 1 \\ 1 & 0 & 0 & 0 & 0 & 0 \\ 0 & 1 & 1 & -1 & -1 & -1 \end{bmatrix} \begin{bmatrix} 3+\theta \\ 15 \\ 10 \\ 6+\theta \\ 10 \\ 5 \end{bmatrix} = \begin{bmatrix} 6+\theta \\ 4-\theta \\ 5+0\theta \\ 6+\theta \\ 3+\theta \\ 4-\theta \end{bmatrix}$$

$$\text{(ii)} \quad B^{-1}R_{13} = \begin{bmatrix} 0 & 0 & 0 & 1 & 0 & 0 \\ 0 & 1 & 0 & -1 & 0 & -1 \\ 0 & 0 & 0 & 0 & 0 & 1 \\ 0 & -1 & 0 & 1 & 1 & 1 \\ 1 & 0 & 0 & 0 & 0 & 0 \\ 0 & 1 & 1 & -1 & -1 & -1 \end{bmatrix} \begin{bmatrix} 3+\theta \\ 15 \\ 10 \\ 6 \\ 10 \\ 5+\theta \end{bmatrix} = \begin{bmatrix} 6+0\theta \\ 4-\theta \\ 5+\theta \\ 6+\theta \\ 3+\theta \\ 4-\theta \end{bmatrix}$$

$$\text{(iii)} \quad B^{-1}R_{31} = \begin{bmatrix} 0 & 0 & 0 & 1 & 0 & 0 \\ 0 & 1 & 0 & -1 & 0 & -1 \\ 0 & 0 & 0 & 0 & 0 & 1 \\ 0 & -1 & 0 & 1 & 1 & 1 \\ 1 & 0 & 0 & 0 & 0 & 0 \\ 0 & 1 & 1 & -1 & -1 & -1 \end{bmatrix} \begin{bmatrix} 3 \\ 15 \\ 10+\theta \\ 6+\theta \\ 10 \\ 5 \end{bmatrix} = \begin{bmatrix} 6+\theta \\ 4-\theta \\ 5+0\theta \\ 6+\theta \\ 3+0\theta \\ 4+0\theta \end{bmatrix}$$

(iv) $$B^{-1}R_{33} = \begin{bmatrix} 0 & 0 & 0 & 1 & 0 & 0 \\ 0 & 1 & 0 & -1 & 0 & -1 \\ 0 & 0 & 0 & 0 & 0 & 1 \\ 0 & -1 & 0 & 1 & 1 & 1 \\ 1 & 0 & 0 & 0 & 0 & 0 \\ 0 & 1 & 1 & -1 & -1 & -1 \end{bmatrix} \begin{bmatrix} 3 \\ 15 \\ 10+\theta \\ 6 \\ 10 \\ 5+\theta \end{bmatrix} = \begin{bmatrix} 6+0\theta \\ 4-\theta \\ 5+\theta \\ 6+\theta \\ 3+0\theta \\ 4+0\theta \end{bmatrix}$$

Step 2.6: According to **Step 2.6** of the proposed Mehar method-III, there is a need to find the range of θ (where, $\theta > 0$) so that all the elements of the matrix, obtained in **Step 2.5**, are greater than or equal to zero.

It is obvious from **Step 2.5** that

(i) The range of θ for which all the elements of the matrix $B^{-1}R_{11}$, obtained in **Step 2.5**, are greater than or equal to zero, i.e., $6+\theta \geq 0, 4-\theta \geq 0, 3+\theta \geq 0$ is $0 < \theta \leq 4$.
(ii) The range of θ for which all the elements of the matrix $B^{-1}R_{13}$, obtained in **Step 2.5**, are greater than or equal to zero, i.e., $4-\theta \geq 0, 5+\theta \geq 0, 6+\theta \geq 0, 3+\theta \geq 0$ is $0 < \theta \leq 4$.
(iii) The range of θ for which all the elements of the matrix $B^{-1}R_{31}$, obtained in **Step 2.5**, are greater than or equal to zero, i.e., $6+\theta \geq 0, 4-\theta \geq 0$ is $0 < \theta \leq 4$.
(iv) The range of θ for which all the elements of the matrix $B^{-1}R_{33}$, obtained in **Step 2.5**, are greater than or equal to zero, i.e., $4-\theta \geq 0, 5+\theta \geq 0, 6+\theta \geq 0$ is $0 < \theta \leq 4$.

Step 2.7: According to **Step 2.7** of the proposed Mehar method-III,

(i) For any value θ such that $0 < \theta \leq 4$,
$6+\theta$ represents the value of the 1st basic variable i.e., x^2_{21},
$4-\theta$ represents the value of the 2nd basic variable i.e., x^2_{22},
$5+0\theta$ represents the value of the 3rd variable i.e., x^2_{23},
$6+\theta$ represents the value of the 4th basic variable i.e., x^2_{32},
$3+\theta$ represents the value of the 5th basic variable i.e., s^2_1 and
$4-\theta$ represents the value of the 6th variable i.e., s^2_2 in a more-for-less solution of crisp transportation problem with mixed constraints represented by Table 4.4.

For example, substituting $\theta = 4$, the obtained values of basic variables in a more-for-less solution are

$$x^2_{21} = 10,$$
$$x^2_{22} = 0,$$
$$x^2_{23} = 5,$$
$$x^2_{32} = 10,$$
$$s^2_1 = 7,$$

$$s_2^2 = 0.$$

(ii) For any value θ such that $0 < \theta \leq 4$,
$6 + 0\theta$ represents the value of the 1st basic variable i.e., x_{21}^2,
$4 - \theta$ represents the value of the 2nd basic variable i.e., x_{22}^2,
$5 + \theta$ represents the value of the 3rd variable i.e., x_{23}^2,
$6 + \theta$ represents the value of the 4th basic variable i.e., x_{32}^2,
$3 + \theta$ represents the value of the 5th basic variable i.e., s_1^2 and
$4 - \theta$ represents the value of the 6th variable i.e., s_2^2 in a more-for-less solution of crisp transportation problem with mixed constraints represented by Table 4.4.

For example, substituting $\theta = 4$, the obtained values of basic variables in a more-for-less solution are

$$x_{21}^2 = 6,$$
$$x_{22}^2 = 0,$$
$$x_{23}^2 = 9,$$
$$x_{32}^2 = 10,$$
$$s_1^2 = 7,$$
$$s_2^2 = 0.$$

(iii) For any value θ such that $0 < \theta \leq 4$,
$6 + \theta$ represents the value of the 1st basic variable i.e., x_{21}^2,
$4 - \theta$ represents the value of the 2nd basic variable i.e., x_{22}^2,
$5 + 0\theta$ represents the value of the 3rd variable i.e., x_{23}^2,
$6 + \theta$ represents the value of the 4th basic variable i.e., x_{32}^2,
$3 + 0\theta$ represents the value of the 5th basic variable i.e., s_1^2 and
$4 + 0\theta$ represents the value of the 6th variable i.e., s_2^2 in a more-for-less solution of crisp transportation problem with mixed constraints represented by Table 4.4.

For example, substituting $\theta = 4$, the obtained values of basic variables in a more-for-less solution are

$$x_{21}^2 = 10,$$
$$x_{22}^2 = 0,$$
$$x_{23}^2 = 5,$$
$$x_{32}^2 = 10,$$
$$s_1^2 = 3,$$
$$s_2^2 = 4.$$

(iv) For any value θ such that $0 < \theta \leq 4$,
$6 + 0\theta$ represents the value of the 1st basic variable i.e., x_{21}^2,
$4 - \theta$ represents the value of the 2nd basic variable i.e., x_{22}^2,
$5 + \theta$ represents the value of the 3rd variable i.e., x_{23}^2,
$6 + \theta$ represents the value of the 4th basic variable i.e., x_{32}^2,
$3 + 0\theta$ represents the value of the 5th basic variable i.e., s_1^2 and
$4 + 0\theta$ represents the value of the 6th variable i.e., s_2^2 in a more-for-less solution of crisp transportation problem with mixed constraints represented by Table 4.4.

For example, substituting $\theta = 4$, the obtained values of basic variables in a more-for-less solution are

$$x_{21}^2 = 6,$$
$$x_{22}^2 = 0,$$
$$x_{23}^2 = 9, x_{32}^2 = 10,$$
$$s_1^2 = 3,$$
$$s_2^2 = 4.$$

Step 2.8: According to **Step 2.8** of the proposed Mehar method-III,

(i) For any value θ such that $0 < \theta \leq 4$,

$$(6 + \theta, \delta_{213}(1 - \alpha), \delta_{213}(1 - \alpha); 6 + \theta, \delta_{214}(1 - \alpha), \delta_{214}(1 - \alpha))$$

represents the value of the 1st basic variable i.e., $\tilde{x}_{21}$,

$$(4 - \theta, \delta_{223}(1 - \alpha), \delta_{223}(1 - \alpha); 4 - \theta, \delta_{224}(1 - \alpha), \delta_{224}(1 - \alpha))$$

represents the value of the 2nd basic variable i.e., $\tilde{x}_{22}$,

$$(5 + 0\theta, \delta_{233}(1 - \alpha), \delta_{233}(1 - \alpha); 5 + 0\theta, \delta_{234}(1 - \alpha), \delta_{234}(1 - \alpha))$$

represents the value of the 3rd variable i.e., $\tilde{x}_{23}$,

$$(6 + \theta, \delta_{323}(1 - \alpha), \delta_{323}(1 - \alpha); 6 + \theta, \delta_{324}(1 - \alpha), \delta_{324}(1 - \alpha))$$

represents the value of the 4th basic variable i.e., $\tilde{x}_{32}$,

$$(3 + \theta, \gamma_{11}(1 - \alpha), \gamma_{11}(1 - \alpha); 3 + \theta, \gamma_{12}(1 - \alpha), \gamma_{12}(1 - \alpha))$$

represents the value of the 5th basic variable i.e., $\tilde{s}_1$ and

$$(4 - \theta, \gamma_{21}(1 - \alpha), \gamma_{21}(1 - \alpha); 4 - \theta, \gamma_{22}(1 - \alpha), \gamma_{22}(1 - \alpha))$$

represents the value of the 6th variable i.e., $\tilde{s}_2$ in a more-for-less solution of the symmetric triangular intuitionistic fuzzy transportation problem with mixed constraints represented by Table 4.3.

For example,

(a) Substituting $\theta = 4$, $\delta_{ij3} = 1, \delta_{ij4} = 2 \forall i, j, \gamma_{k1} = \gamma_{k2} = 1; k = 1, 2$, the obtained values of basic variables in a more-for-less solution are

$$\begin{aligned}
\tilde{x}_{21} &=_{\text{Mag}} (10, 1(1-\alpha), 1(1-\alpha); 10, 2(1-\alpha), 2(1-\alpha)),\\
\tilde{x}_{22} &=_{\text{Mag}} (0, 1(1-\alpha), 1(1-\alpha); 0, 2(1-\alpha), 2(1-\alpha)),\\
\tilde{x}_{23} &=_{\text{Mag}} (5, 1(1-\alpha), 1(1-\alpha); 5, 2(1-\alpha), 2(1-\alpha)),\\
\tilde{x}_{32} &=_{\text{Mag}} (10, 1(1-\alpha), 1(1-\alpha); 10, 2(1-\alpha), 2(1-\alpha)),\\
\tilde{s}_1 &=_{\text{Mag}} (7, 1(1-\alpha), 1(1-\alpha); 7, 1(1-\alpha), 1(1-\alpha)),\\
\tilde{s}_2 &=_{\text{Mag}} (0, 1(1-\alpha), 1(1-\alpha); 0, 1(1-\alpha), 1(1-\alpha))
\end{aligned}$$

and the corresponding symmetric triangular intuitionistic fuzzy transportation cost is $\tilde{Z} =_{\text{Mag}} (90, 2(1-\alpha), 2(1-\alpha); 90, 3(1-\alpha), 3(1-\alpha))$.

(b) Substituting $\theta = 4$, $\delta_{ij3} = 4, \delta_{ij4} = 7 \forall i, j, \gamma_{k1} = \gamma_{k2} = 1; k = 1, 2$, the obtained values of basic variables in a more-for-less solution are

$$\begin{aligned}
\tilde{x}_{21} &=_{\text{Mag}} (10, 4(1-\alpha), 4(1-\alpha); 10, 7(1-\alpha), 7(1-\alpha)),\\
\tilde{x}_{22} &=_{\text{Mag}} (0, 4(1-\alpha), 4(1-\alpha); 0, 7(1-\alpha), 7(1-\alpha)),\\
\tilde{x}_{23} &=_{\text{Mag}} (5, 4(1-\alpha), 4(1-\alpha); 5, 7(1-\alpha), 7(1-\alpha)),\\
\tilde{x}_{32} &=_{\text{Mag}} (10, 4(1-\alpha), 4(1-\alpha); 10, 7(1-\alpha), 7(1-\alpha)),\\
\tilde{s}_1 &=_{\text{Mag}} (7, 1(1-\alpha), 1(1-\alpha); 7, 1(1-\alpha), 1(1-\alpha)),\\
\tilde{s}_2 &=_{\text{Mag}} (0, 1(1-\alpha), 1(1-\alpha); 0, 1(1-\alpha), 1(1-\alpha))
\end{aligned}$$

and the corresponding symmetric triangular intuitionistic fuzzy transportation cost is $\tilde{Z} =_{\text{Mag}} (90, 4(1-\alpha), 4(1-\alpha); 90, 7(1-\alpha), 7(1-\alpha))$.

(ii) For any value θ such that $0 < \theta \leq 4$,

$$(6 + 0\theta, \delta_{213}(1-\alpha), \delta_{213}(1-\alpha); 6 + 0\theta, \delta_{214}(1-\alpha), \delta_{214}(1-\alpha))$$

represents the value of the 1st basic variable i.e., $\tilde{x}_{21}$,

$$(4 - \theta, \delta_{223}(1-\alpha), \delta_{223}(1-\alpha); 4 - \theta, \delta_{224}(1-\alpha), \delta_{224}(1-\alpha))$$

represents the value of the 2nd basic variable i.e., $\tilde{x}_{22}$,

$$(5 + \theta, \delta_{233}(1-\alpha), \delta_{233}(1-\alpha); 5 + \theta, \delta_{234}(1-\alpha), \delta_{234}(1-\alpha))$$

represents the value of the 3rd variable i.e., $\tilde{x}_{23}$,

$$(6+\theta, \delta_{323}(1-\alpha), \delta_{323}(1-\alpha); 6+\theta, \delta_{324}(1-\alpha), \delta_{324}(1-\alpha))$$

represents the value of the 4th basic variable i.e., $\tilde{x}_{32}$,

$$(3+\theta, \gamma_{11}(1-\alpha), \gamma_{11}(1-\alpha); 3+\theta, \gamma_{12}(1-\alpha), \gamma_{12}(1-\alpha))$$

represents the value of the 5th basic variable i.e., $\tilde{s}_1$ and

$$(4-\theta, \gamma_{21}(1-\alpha), \gamma_{21}(1-\alpha); 4-\theta, \gamma_{22}(1-\alpha), \gamma_{22}(1-\alpha))$$

represents the value of the 6th variable i.e., $\tilde{s}_2$ in a more-for-less solution of the symmetric triangular intuitionistic fuzzy transportation problem with mixed constraints represented by Table 4.3.

For example,

(a) Substituting $\theta = 4, \delta_{ij3} = 3, \delta_{ij4} = 4 \forall i, j, \gamma_{k1} = \gamma_{k2} = 1; k = 1, 2$, the obtained values of basic variables in a more-for-less solution are

$$\begin{aligned}
\tilde{x}_{21} &=_{\text{Mag}} (6, 3(1-\alpha), 3(1-\alpha); 6, 4(1-\alpha), 4(1-\alpha)),\\
\tilde{x}_{21} &=_{\text{Mag}} (0, 3(1-\alpha), 3(1-\alpha); 0, 4(1-\alpha), 4(1-\alpha)),\\
\tilde{x}_{23} &=_{\text{Mag}} (9, 3(1-\alpha), 3(1-\alpha); 9, 4(1-\alpha), 4(1-\alpha)),\\
\tilde{x}_{32} &=_{\text{Mag}} (10, 3(1-\alpha), 3(1-\alpha); 10, 4(1-\alpha), 4(1-\alpha)),\\
\tilde{s}_1 &=_{\text{Mag}} (7, 1(1-\alpha), 1(1-\alpha); 7, 1(1-\alpha), 1(1-\alpha)),\\
\tilde{s}_2 &=_{\text{Mag}} (0, 1(1-\alpha), 1(1-\alpha); 0, 1(1-\alpha), 1(1-\alpha))
\end{aligned}$$

and the corresponding symmetric triangular intuitionistic fuzzy transportation cost is $\tilde{Z} =_{\text{Mag}} (86, 3(1-\alpha), 3(1-\alpha); 86, 4(1-\alpha), 4(1-\alpha))$.

(b) Substituting $\theta = 4, \delta_{ij3} = 1, \delta_{ij4} = 4 \forall i, j, \gamma_{k1} = \gamma_{k2} = 1; k = 1, 2$, the obtained values of basic variables in a more-for-less solution are

$$\begin{aligned}
\tilde{x}_{21} &=_{\text{Mag}} (6, 1(1-\alpha), 1(1-\alpha); 6, 4(1-\alpha), 4(1-\alpha)),\\
\tilde{x}_{21} &=_{\text{Mag}} (0, 1(1-\alpha), 1(1-\alpha); 0, 4(1-\alpha), 4(1-\alpha)),\\
\tilde{x}_{23} &=_{\text{Mag}} (9, 1(1-\alpha), 1(1-\alpha); 9, 4(1-\alpha), 4(1-\alpha)),\\
\tilde{x}_{32} &=_{\text{Mag}} (10, 1(1-\alpha), 1(1-\alpha); 10, 4(1-\alpha), 4(1-\alpha)),\\
\tilde{s}_1 &=_{\text{Mag}} (7, 1(1-\alpha), 1(1-\alpha); 7, 1(1-\alpha), 1(1-\alpha)),\\
\tilde{s}_2 &=_{\text{Mag}} (0, 1(1-\alpha), 1(1-\alpha); 0, 1(1-\alpha), 1(1-\alpha))
\end{aligned}$$

and the corresponding symmetric triangular intuitionistic fuzzy transportation cost is $\tilde{Z} =_{\text{Mag}} (86, 2(1-\alpha), 2(1-\alpha); 86, 4(1-\alpha), 4(1-\alpha))$.

(iii) For any value θ such that $0 < \theta \leq 4$,

$$(6+\theta, \delta_{213}(1-\alpha), \delta_{213}(1-\alpha); 6+\theta, \delta_{214}(1-\alpha), \delta_{214}(1-\alpha))$$

represents the value of the 1st basic variable i.e., $\tilde{x}_{21}$,

$$(4-\theta, \delta_{223}(1-\alpha), \delta_{223}(1-\alpha); 4-\theta, \delta_{224}(1-\alpha), \delta_{224}(1-\alpha))$$

represents the value of the 2nd basic variable i.e., $\tilde{x}_{22}$,

$$(5+0\theta, \delta_{233}(1-\alpha), \delta_{233}(1-\alpha); 5+0\theta, \delta_{234}(1-\alpha), \delta_{234}(1-\alpha))$$

represents the value of the 3rd variable i.e., $\tilde{x}_{23}$,

$$(6+\theta, \delta_{323}(1-\alpha), \delta_{323}(1-\alpha); 6+\theta, \delta_{324}(1-\alpha), \delta_{324}(1-\alpha))$$

represents the value of the 4th basic variable i.e., $\tilde{x}_{32}$,

$$(3+0\theta, \gamma_{11}(1-\alpha), \gamma_{11}(1-\alpha); 3+0\theta, \gamma_{12}(1-\alpha), \gamma_{12}(1-\alpha))$$

represents the value of the 5th basic variable i.e., $\tilde{s}_1$ and

$$(4+0\theta, \gamma_{21}(1-\alpha), \gamma_{21}(1-\alpha); 4+0\theta, \gamma_{22}(1-\alpha), \gamma_{22}(1-\alpha))$$

represents the value of the 6th variable i.e., $\tilde{s}_2$ in a more-for-less solution of the symmetric triangular intuitionistic fuzzy transportation problem with mixed constraints represented by Table 4.3.

For example,

(a) Substituting $\theta = 4, \delta_{ij3} = 1, \delta_{ij4} = 5 \forall i, j, \gamma_{k1} = \gamma_{k2} = 1; k = 1, 2$, the obtained values of basic variables in a more-for-less solution are $\tilde{x}_{21} =_{\text{Mag}} (10, 1(1-\alpha), 1(1-\alpha); 10, 5(1-\alpha), 5(1-\alpha))$,
$\tilde{x}_{22} =_{\text{Mag}} (0, 1(1-\alpha), 1(1-\alpha); 0, 5(1-\alpha), 5(1-\alpha))$,
$\tilde{x}_{23} =_{\text{Mag}} (5, 1(1-\alpha), 1(1-\alpha); 5, 5(1-\alpha), 5(1-\alpha))$,
$\tilde{x}_{32} =_{\text{Mag}} (10, 1(1-\alpha), 1(1-\alpha); 10, 5(1-\alpha), 5(1-\alpha))$,
$\tilde{s}_1 =_{\text{Mag}} (3, 1(1-\alpha), 1(1-\alpha); 3, 1(1-\alpha), 1(1-\alpha))$,
$\tilde{s}_2 =_{\text{Mag}} (4, 1(1-\alpha), 1(1-\alpha); 4, 1(1-\alpha), 1(1-\alpha))$

and the corresponding symmetric triangular intuitionistic fuzzy transportation cost is $\tilde{Z} =_{\text{Mag}} (90, 2(1-\alpha), 2(1-\alpha); 90, 5(1-\alpha), 5(1-\alpha))$.

(b) Substituting $\theta = 4, \delta_{ij3} = \delta_{ij4} = 2 \forall i, j, \gamma_{k1} = \gamma_{k2} = 1; k = 1, 2$, the obtained values of basic variables in a more-for-less solution are $\tilde{x}_{21} =_{\text{Mag}} (10, 2(1-\alpha), 2(1-\alpha); 10, 2(1-\alpha), 2(1-\alpha))$,
$\tilde{x}_{22} =_{\text{Mag}} (0, 2(1-\alpha), 2(1-\alpha); 0, 2(1-\alpha), 2(1-\alpha))$,
$\tilde{x}_{23} =_{\text{Mag}} (5, 2(1-\alpha), 2(1-\alpha); 5, 2(1-\alpha), 2(1-\alpha))$,
$\tilde{x}_{32} =_{\text{Mag}} (10, 2(1-\alpha), 2(1-\alpha); 10, 2(1-\alpha), 2(1-\alpha))$,
$\tilde{s}_1 =_{\text{Mag}} (3, 1(1-\alpha), 1(1-\alpha); 3, 1(1-\alpha), 1(1-\alpha))$,
$\tilde{s}_2 =_{\text{Mag}} (4, 1(1-\alpha), 1(1-\alpha); 4, 1(1-\alpha), 1(1-\alpha))$

and the corresponding symmetric triangular intuitionistic fuzzy transportation cost is $\tilde{Z} =_{\text{Mag}} (90, 2(1-\alpha), 2(1-\alpha); 90, 3(1-\alpha), 3(1-\alpha))$.

(iv) For any value θ such that $0 < \theta \leq 4$,

$$(6 + 0\theta, \delta_{213}(1-\alpha), \delta_{213}(1-\alpha); 6 + 0\theta, \delta_{214}(1-\alpha), \delta_{214}(1-\alpha))$$

represents the value of the 1st basic variable i.e., $\tilde{x}_{21}$,

$$(4 - \theta, \delta_{223}(1-\alpha), \delta_{223}(1-\alpha); 4 - \theta, \delta_{224}(1-\alpha), \delta_{224}(1-\alpha))$$

represents the value of the 2nd basic variable i.e., $\tilde{x}_{22}$,

$$(5 + \theta, \delta_{233}(1-\alpha), \delta_{233}(1-\alpha); 5 + \theta, \delta_{234}(1-\alpha), \delta_{234}(1-\alpha))$$

represents the value of the 3rd variable i.e., $\tilde{x}_{23}$,

$$(6 + \theta, \delta_{323}(1-\alpha), \delta_{323}(1-\alpha); 6 + \theta, \delta_{324}(1-\alpha), \delta_{324}(1-\alpha))$$

represents the value of the 4th basic variable i.e., $\tilde{x}_{32}$,

$$(3 + 0\theta, \gamma_{11}(1-\alpha), \gamma_{11}(1-\alpha); 3 + 0\theta, \gamma_{12}(1-\alpha), \gamma_{12}(1-\alpha))$$

represents the value of the 5th basic variable i.e., $\tilde{s}_1$ and

$$(4 + 0\theta, \gamma_{21}(1-\alpha), \gamma_{21}(1-\alpha); 4 + 0\theta, \gamma_{22}(1-\alpha), \gamma_{22}(1-\alpha))$$

represents the value of the 6th variable i.e., $\tilde{s}_2$ in a more-for-less solution of the symmetric triangular intuitionistic fuzzy transportation problem with mixed constraints represented by Table 4.3.

For example,

(a) Substituting $\theta = 4$, $\delta_{ij3} = \delta_{ij4} = 1 \forall i, j$, $\gamma_{k1} = \gamma_{k2} = 1; k = 1, 2$, the obtained values of basic variables in a more-for-less solution are
$\tilde{x}_{21} =_{\text{Mag}} (6, 1(1-\alpha), 1(1-\alpha); 6, 1(1-\alpha), 1(1-\alpha))$,
$\tilde{x}_{22} =_{\text{Mag}} (0, 1(1-\alpha), 1(1-\alpha); 0, 1(1-\alpha), 1(1-\alpha))$,
$\tilde{x}_{23} =_{\text{Mag}} (9, 1(1-\alpha), 1(1-\alpha); 9, 1(1-\alpha), 1(1-\alpha))$,
$\tilde{x}_{32} =_{\text{Mag}} (10, 1(1-\alpha), 1(1-\alpha); 10, 1(1-\alpha), 1(1-\alpha))$,
$\tilde{s}_1 =_{\text{Mag}} (3, 1(1-\alpha), 1(1-\alpha); 3, 1(1-\alpha), 1(1-\alpha))$,
$\tilde{s}_2 =_{\text{Mag}} (4, 1(1-\alpha), 1(1-\alpha); 4, 1(1-\alpha), 1(1-\alpha))$

and the corresponding symmetric triangular intuitionistic fuzzy transportation cost is $\tilde{Z} =_{\text{Mag}} (86, 2(1-\alpha), 2(1-\alpha); 86, 3(1-\alpha), 3(1-\alpha))$.

(b) Substituting $\theta = 4, \delta_{ij3} = 1, \delta_{ij4} = 6 \forall i, j, \gamma_{k1} = \gamma_{k2} = 1; k = 1, 2,$ the obtained values of basic variables in a more-for-less solution are

$$\tilde{x}_{21} =_{\text{Mag}} (6, 1(1-\alpha), 1(1-\alpha); 6, 6(1-\alpha), 6(1-\alpha)),$$
$$\tilde{x}_{22} =_{\text{Mag}} (0, 1(1-\alpha), 1(1-\alpha); 0, 6(1-\alpha), 6(1-\alpha)),$$
$$\tilde{x}_{23} =_{\text{Mag}} (9, 1(1-\alpha), 1(1-\alpha); 9, 6(1-\alpha), 6(1-\alpha)),$$
$$\tilde{x}_{32} =_{\text{Mag}} (10, 1(1-\alpha), 1(1-\alpha); 10, 6(1-\alpha), 6(1-\alpha)),$$
$$\tilde{s}_{1} =_{\text{Mag}} (3, 1(1-\alpha), 1(1-\alpha); 3, 1(1-\alpha), 1(1-\alpha)),$$
$$\tilde{s}_{2} =_{\text{Mag}} (4, 1(1-\alpha), 1(1-\alpha); 4, 1(1-\alpha), 1(1-\alpha))$$

and the corresponding symmetric triangular intuitionistic fuzzy transportation cost is $\tilde{Z} =_{\text{Mag}} (86, 2(1-\alpha), 2(1-\alpha); 86, 6(1-\alpha), 6(1-\alpha))$.

4.11 Conclusions

In this chapter, new arithmetic operations of triangular intuitionistic fuzzy numbers are proposed. Also, a new method for comparing two triangular intuitionistic fuzzy numbers is proposed. Finally, a new method (named as Mehar method-III) is proposed to find more-for-less solutions of symmetric intuitionistic fuzzy transportation problems with mixed constraints.

References

1. A. Mishra and A. Kumar, Aggregation operators for various extensions of fuzzy set and its applications in transportation problems, Studies in Fuzziness and Soft Computing, Springer Nature, Singapore, (2021).
2. H. A. Taha, Operations research: An introduction, Pearson Prentice Hall Upper Saddle River, N.J. (2013).
3. V. Vidhya, P. Uma Maheswari and K. Ganesan, An alternate method for finding more for less solution to fuzzy transportation problem with mixed constraints, Soft Computing 25 (2021) 11989–11996.

Chapter 5
Mehar Method-IV to Find All More-For-Less Solutions of Symmetric Intuitionistic Fuzzy Linear Fractional Transportation Problems with Mixed Constraints

In this chapter, a new method (named as Mehar method-IV) is proposed to check the existence of at least one more-for-less solution as well as to find all more-for-less solutions (if exists) of symmetric triangular intuitionistic fuzzy linear fractional transportation problems with mixed constraints (linear fractional transportation problems with mixed constraints in which each parameter and each variable is represented by a symmetric triangular intuitionistic fuzzy number). Also, to illustrate the proposed Mehar method-IV, an existing crisp linear fractional transportation problem with mixed constraints and a symmetric triangular intuitionistic fuzzy linear fractional transportation problem with mixed constraints are solved by the proposed Mehar method-IV. Furthermore, the superiority of the proposed Mehar method-IV over an existing method is discussed.

5.1 Tabular Representation of Crisp Linear Fractional Transportation Problems with Mixed Constraints

A crisp linear fractional transportation problem with mixed constraints can be represented into tabular form as shown by Table 5.1.

5.2 Tabular Representation of Symmetric Triangular Intuitionistic Fuzzy Linear Fractional Transportation Problems with Mixed Constraints

A symmetric triangular intuitionistic fuzzy linear fractional transportation problem with mixed constraints can be represented into tabular form as shown by Table 5.2.

T. K. Bhatia et al., *More-for-Less Solutions in Fuzzy Transportation Problems*, Studies in Fuzziness and Soft Computing 426,
https://doi.org/10.1007/978-3-031-30337-1_5

Table 5.1 Tabular representation of a crisp linear fractional transportation problem with mixed constraints

	D_1	D_2	...	D_n	Availability
S_1	c_{11} d_{11}	c_{12} d_{12}	...	c_{1n} d_{1n}	$=$ or $\leq$ or $\geq a_1$
S_2	c_{21} d_{21}	c_{22} d_{22}	...	c_{2n} d_{2n}	$=$ or $\leq$ or $\geq a_2$
$\vdots$	$\vdots$	$\vdots$	$\vdots$	$\vdots$	$\vdots$
S_m	c_{m1} d_{m1}	c_{m2} d_{m2}	...	c_{mn} d_{mn}	$=$ or $\leq$ or $\geq a_m$
Demand	$=$ or $\leq$ or $\geq b_1$	$=$ or $\leq$ or $\geq b_2$	...	$=$ or $\leq$ or $\geq b_n$	

where,
(i) m represents the number of sources
(ii) n represents the number of destinations
(iii) S_i represents the ith source
(iv) D_j represents the jth destination
(v) The positive real number c_{ij} represents the cost for transporting one unit quantity of the product from the ith source to the jth destination
(vi) The positive real number d_{ij} represents the profit on transporting one unit quantity of the product from the ith source to the jth destination
(vii) The positive real number a_i represents the availability of the product at the ith source
(viii) The positive real number b_j represents the demand of the product at the jth destination

5.3 Crisp Linear Fractional Programming Problems Corresponding to Crisp Linear Fractional Transportation Problem with Mixed Constraints

If it is assumed that the non-negative real number x_{ij} represents the quantity of the product to be supplied from the ith source to the jth destination and Z represents the total crisp transportation cost. Then, the crisp linear fractional transportation problem with mixed constraints, represented by Table 5.1, can be formulated into its equivalent crisp linear fractional programming problem ($P5.1$) or into its equivalent crisp linear programming problem ($P5.2$) (dual of the crisp linear fractional programming problem ($P5.1$)).

Problem (P5.1) *Minimize* $\left(Z = \frac{\sum_{i=1}^{m}\sum_{j=1}^{n} c_{ij}x_{ij}}{\sum_{i=1}^{m}\sum_{j=1}^{n} d_{ij}x_{ij}}\right)$
Subject to

$$\sum_{j=1}^{n} x_{ij} = or \geq or \leq a_i;\ i = 1, 2, \ldots, m;\ (i\text{th source constraint})$$

$$\sum_{i=1}^{m} x_{ij} = or \geq or \leq b_j;\ j = 1, 2, \ldots, n;\ (j\text{th destination constraint})$$

Table 5.2 Tabular representation of a symmetric triangular intuitionistic fuzzy linear fractional transportation problem with mixed constraints

	D_1	D_2	…	D_n	Availability
S_1	$\tilde{c}_{11}$ $\tilde{d}_{11}$	$\tilde{c}_{12}$ $\tilde{d}_{12}$	…	$\tilde{c}_{1n}$ $\tilde{d}_{1n}$	$=_{Mag}$ *or* $\leq_{Mag}$ *or* $\geq_{Mag}$ $\tilde{a}_1$
S_2	$\tilde{c}_{21}$ $\tilde{d}_{21}$	$\tilde{c}_{22}$ $\tilde{d}_{22}$	…	$\tilde{c}_{2n}$ $\tilde{d}_{2n}$	$=_{Mag}$ *or* $\leq_{Mag}$ *or* $\geq_{Mag}$ $\tilde{a}_2$
⋮	⋮	⋮	⋮	⋮	⋮
S_m	$\tilde{c}_{m1}$ $\tilde{d}_{m1}$	$\tilde{c}_{m2}$ $\tilde{d}_{m1}$	…	$\tilde{c}_{mn}$ $\tilde{d}_{mn}$	$=_{Mag}$ *or* $\leq_{Mag}$ *or* $\geq_{Mag}$ $\tilde{a}_m$
Demand	$=_{Mag}$ *or* $\leq_{Mag}$ *or* $\geq_{Mag}$ $\tilde{b}_1$	$=_{Mag}$ *or* $\leq_{Mag}$ *or* $\geq_{Mag}$ $\tilde{b}_2$	…	$=_{Mag}$ *or* $\leq_{Mag}$ *or* $\geq_{Mag}$ $\tilde{b}_n$	

where,

(i) The symmetric triangular intuitionistic fuzzy number $\tilde{a}_i = (a_i^2, \delta_{i1}(1-\alpha), \delta_{i1}(1-\alpha); a_i^2, \delta_{i2}(1-\alpha), \delta_{i2}(1-\alpha))$ represents the supply of product at the ith source

(ii) The symmetric triangular intuitionistic fuzzy number $\tilde{b}_j = \left(b_j^2, \delta_{j3}(1-\alpha), \delta_{j3}(1-\alpha); b_j^2, \delta_{j4}(1-\alpha), \delta_{j4}(1-\alpha)\right)$ represents the demand of product at the jth destination

(iii) The symmetric triangular intuitionistic fuzzy number $\tilde{c}_{ij} = \left(c_{ij}^2, \delta_{ij1}(1-\alpha), \delta_{ij1}(1-\alpha); c_{ij}^2, \delta_{ij2}(1-\alpha), \delta_{ij2}(1-\alpha)\right)$ represents the cost for transporting one unit quantity of product from the ith source to the jth destination

(iv) The symmetric triangular intuitionistic fuzzy number $\tilde{d}_{ij} = \left(d_{ij}^2, \delta_{ij3}(1-\alpha), \delta_{ij3}(1-\alpha); d_{ij}^2, \delta_{ij4}(1-\alpha), \delta_{ij4}(1-\alpha)\right)$ represents the profit on transporting one unit quantity of the product from the ith source to the jth destination

$$x_{ij} \geq 0; i = 1, 2, \ldots, m; j = 1, 2, \ldots n,$$

where,

$\sum_{i=1}^{m}\sum_{j=1}^{n} d_{ij}x_{ij} > 0 \forall x_{ij} \in S$, S represents the set of all feasible solutions of the problem $(P5.1)$.

Problem $(\boldsymbol{P5.2})$ *Maximize* (k)

Subject to

$$-\sum_{i=1}^{m} a_i u_i - \sum_{j=1}^{n} b_j v_j - \sum_{i=1}^{m}\sum_{j=1}^{n} P_{ij} w_{ij} \leq 0,$$

$$u_i + v_j + Q_{ij} w_{ij} + d_{ij} k \leq c_{ij}; i = 1, 2, \ldots, m; j = 1, 2, \ldots, n,$$

$$u_i * 0, v_j \circ 0; i = 1, 2, \ldots, m; j = 1, 2, \ldots, n,$$

$$w_{ij} \leq 0,$$

k is unrestricted dual variable,

where,

(i) $P_{ij} = \begin{cases} M; & \begin{array}{l} if\ on\ solving\ the\ problem\left(P5.1^{ij1}\right) \\ an\ unbounded\ optimal\ solution\ is\ obtained \end{array} \\ 0; & \begin{array}{l} if\ on\ solving\ the\ problem\left(P5.1^{ij1}\right) \\ a\ bounded\ optimal\ solution\ is\ obtained \end{array} \end{cases}$

where, M is a very large positive real number tends to $+\infty$.

Problem $\left(\boldsymbol{P}5.1^{ij1}\right)$

$Maximize\left(x_{ij}\right)$

Subject to

Constraints of the problem $(P5.1)$.

(ii) $Q_{ij} = \begin{cases} 1; & \begin{array}{l} if\ on\ solving\ the\ problem\left(P5.1^{ij1}\right) \\ an\ unbounded\ optimal\ solution\ is\ obtained \end{array} \\ 0; & \begin{array}{l} if\ on\ solving\ the\ problem\left(P5.1^{ij1}\right) \\ a\ bounded\ optimal\ solution\ is\ obtained \end{array} \end{cases}$

(iii) $*$ will be $\geq$ if the sign of the ith source constraint of the crisp linear fractional programming problem $(P5.1)$ is $\geq$.

(iv) $*$ will be $\leq$ if the sign of the ith source constraint of the crisp linear fractional programming problem $(P5.1)$ is $\leq$.

(v) If the sign of the ith source constraint of the crisp linear fractional programming problem $(P5.1)$ is $=$. Then, u_i will be an unrestricted variable.

(vi) $\circ$ will be $\geq$ if the sign of the jth destination constraint of the crisp linear fractional programming problem $(P5.1)$ is $\geq$.

(vii) $\circ$ will be $\leq$ if the sign of the jth destination constraint of the crisp linear fractional programming problem ($P5.1$) is $\leq$.

(viii) If the sign of the jth destination constraint of the crisp linear fractional programming problem ($P5.1$) $=$. Then, v_j will be an unrestricted variable.

The dual problem ($P5.2$) is obtained as follows.

Step 1: Transform the crisp linear fractional programming problem ($P5.1$) into its equivalent crisp linear fractional programming problem ($P5.1'$).

Problem $(P5.1')$

Minimize $\left(Z = \frac{\sum_{i=1}^{m}\sum_{j=1}^{n} c_{ij}x_{ij}}{\sum_{i=1}^{m}\sum_{j=1}^{n} d_{ij}x_{ij}}\right)$

Subject to

Constraints of the problem ($P5.1$) with the additional constraints

$x_{pq} \leq M$ for all those p, q corresponding to which there exists an unbounded optimal solution of the problem ($P5.1^{pq1}$), where M is a very large positive real number tends to $+\infty$.

Problem $(P5.1^{pq1})$

Maximize (x_{pq})

Subject to

Constraints of the problem ($P5.1$).

Step 2: Using the existing Charnes and Cooper's algorithm [2], transform the crisp linear fractional programming problem ($P5.1$) into its equivalent crisp linear programming problem ($P5.3$).

Problem (**$P5.3$**)

Minimize $\left(\sum_{i=1}^{m}\sum_{j=1}^{n} c_{ij}y_{ij}\right)$

Subject to

$$\sum_{i=1}^{m}\sum_{j=1}^{n} d_{ij}y_{ij} = 1,$$

$$\sum_{j=1}^{n} y_{ij} - a_i t = or \geq or \leq 0;\ i = 1, 2, \ldots, m;$$

$$\sum_{i=1}^{n} y_{ij} - b_j t = or \geq or \leq 0;\ j = 1, 2, \ldots, n;$$

$$y_{ij} - Mt \leq 0,$$

$$y_{ij} \geq 0 \forall i, j,$$

$$t \geq 0.$$

Step 3: Using the duality theory [5], the problem ($P5.2$) represents the dual of the crisp linear programming problem ($P5.3$).

5.4 Intuitionistic Fuzzy Linear Fractional Programming Problems Corresponding to Symmetric Triangular Intuitionistic Fuzzy Linear Fractional Transportation Problem with Mixed Constraints

If it is assumed that the symmetric triangular intuitionistic fuzzy number $\tilde{x}_{ij} = \left(x_{ij}^2, \delta_{ij5}(1-\alpha), \delta_{ij5}(1-\alpha); x_{ij}^2, \delta_{ij6}(1-\alpha), \delta_{ij6}(1-\alpha)\right)$ represents the intuitionistic fuzzy quantity of the product to be supplied from the ith source to the jth destination, the symmetric triangular intuitionistic fuzzy number $\tilde{Z}$ represents the total intuitionistic fuzzy transportation cost and $\tilde{0}$ represents the zero triangular intuitionistic fuzzy number. Then, the symmetric triangular intuitionistic fuzzy linear fractional transportation problem with mixed constraints, represented by Table 5.2, can be formulated into its equivalent intuitionistic fuzzy linear fractional programming problem ($P5.4$) or into its equivalent crisp linear programming problem ($P5.5$) (dual problem of the intuitionistic fuzzy linear fractional programming problem ($P5.4$)).

Problem ($P5.4$) *Minimize* $\left(\tilde{Z} =_{Mag} \frac{\oplus_{i=1}^{m}\oplus_{j=1}^{n}(\tilde{c}_{ij}\otimes\tilde{x}_{ij})}{\oplus_{i=1}^{m}\oplus_{j=1}^{n}(\tilde{d}_{ij}\otimes\tilde{x}_{ij})}\right)$

Subject to

$$\oplus_{j=1}^{n}\tilde{x}_{ij} =_{Mag} or \geq_{Mag} or \leq_{Mag} \tilde{a}_i; i = 1, 2, \ldots, m;$$

$$\oplus_{i=1}^{m}\tilde{x}_{ij} =_{Mag} or \geq_{Mag} or \leq_{Mag} \tilde{b}_j; j = 1, 2, \ldots, n;$$

$$\tilde{x}_{ij} \geq_{Mag} \tilde{0}; i = 1, 2, \ldots, m; j = 1, 2, \ldots, n.$$

Problem ($P5.5$) *Maximize* (k)

Subject to

$$-\sum_{i=1}^{m} a_i^2 u_i^2 - \sum_{j=1}^{n} b_j^2 v_j^2 - \sum_{i=1}^{m}\sum_{j=1}^{n} P_{ij} w_{ij}^2 \leq 0,$$

$$u_i^2 + v_j^2 + Q_{ij} w_{ij}^2 + d_{ij}^2 k \leq c_{ij}^2; i = 1, 2, \ldots, m; j = 1, 2, \ldots, n;$$

$$u_i^2 * 0, v_j^2 \circ 0; i = 1, 2, \ldots, m; j = 1, 2, \ldots, n;$$

$$w_{ij}^2 \leq 0,$$

k is unrestricted dual variable,

where,

(i) $$P_{ij} = \begin{cases} M; & \begin{array}{l} if\ on\ solving\ the\ problem\left(P5.1^{ij2}\right) \\ an\ unbounded\ optimal\ solution\ is\ obtained \end{array} \\ 0; & \begin{array}{l} if\ on\ solving\ the\ problem\left(P5.1^{ij2}\right) \\ a\ bounded\ optimal\ solution\ is\ obtained \end{array} \end{cases}$$

Problem $\left(\boldsymbol{P}5.1^{ij2}\right)$

Maximize $\left(x_{ij}^2\right)$

Subject to

$$\sum_{j=1}^{n} x_{ij}^2 = or \geq or \leq a_i^2;\ i = 1, 2, \ldots, m;$$

$$\sum_{i=1}^{m} x_{ij}^2 = or \geq or \leq b_j^2;\ j = 1, 2, \ldots, n;$$

$$x_{ij}^2 \geq 0;\ i = 1, 2, \ldots, m;\ j = 1, 2, \ldots, n.$$

(ii) $$Q_{ij} = \begin{cases} 1; & \begin{array}{l} if\ on\ solving\ the\ problem\left(P5.1^{ij2}\right) \\ an\ unbounded\ optimal\ solution\ is\ obtained \end{array} \\ 0; & \begin{array}{l} if\ on\ solving\ the\ problem\left(P5.1^{ij2}\right) \\ a\ bounded\ optimal\ solution\ is\ obtained \end{array} \end{cases}$$

(iii) $*$ will be $\geq$ if the sign of the ith source constraint of the intuitionistic fuzzy linear fractional programming problem ($P5.4$) is $\geq$.

(iv) $*$ will be $\leq$ if the sign of the ith source constraint of the intuitionistic fuzzy linear fractional programming problem ($P5.4$) is $\leq$.

(v) If the sign of the ith source constraint of the intuitionistic fuzzy linear fractional programming problem ($P5.4$) is $=$. Then, u_i will be an unrestricted variable.

(vi) $\circ$ will be $\geq$ if the sign of the jth destination constraint of the intuitionistic fuzzy linear fractional programming problem ($P5.4$) is $\geq$.

(vii) $\circ$ will be $\leq$ if the sign of the jth destination constraint of the intuitionistic fuzzy linear fractional programming problem ($P5.4$) is $\leq$.

(viii) If the sign of the jth destination constraint of the intuitionistic fuzzy linear fractional programming problem ($P5.4$) $=$. Then, v_j will be an unrestricted variable.

The dual problem ($P5.5$) is obtained as follows.

Step 1: Using the comparing method, proposed in Sect. 4.3 of Chap. 4, the intuitionistic fuzzy linear fractional programming problem ($P5.4$) can be transformed into its equivalent crisp linear fractional programming problem ($P5.6$).

Problem (*P*5.6)

Minimize $\left(Mag\left(\tilde{Z}\right) = Mag\left(\frac{\oplus_{i=1}^{m}\oplus_{j=1}^{n}\left(\tilde{c}_{ij}\otimes\tilde{x}_{ij}\right)}{\oplus_{i=1}^{m}\oplus_{j=1}^{n}\left(\tilde{d}_{ij}\otimes\tilde{x}_{ij}\right)}\right)\right)$

Subject to

$$Mag\left(\oplus_{j=1}^{n}\tilde{x}_{ij}\right) = or \geq or \leq Mag(\tilde{a}_i);\ i = 1, 2, \ldots, m;$$

$$Mag\left(\oplus_{i=1}^{m}\tilde{x}_{ij}\right) = or \geq or \leq Mag\left(\tilde{b}_j\right);\ j = 1, 2, \ldots, n;$$

$$Mag\left(\tilde{x}_{ij}\right) \geq Mag\left(\tilde{0}\right);\ i = 1, 2, \ldots, m;\ j = 1, 2, \ldots, n.$$

Step 2: Using the expression $Mag\left(\frac{\tilde{A}_1}{\tilde{A}_2}\right) = \frac{Mag\left(\tilde{A}_1\right)}{Mag\left(\tilde{A}_2\right)}$, discussed in Chap. 4, the crisp linear fractional programming problem (*P*5.6) can be transformed into its equivalent crisp linear fractional programming problem (*P*5.7).

Problem (*P*5.7)

Minimize $\left(Mag\left(\tilde{Z}\right) = \frac{Mag\left(\oplus_{i=1}^{m}\oplus_{j=1}^{n}\left(\tilde{c}_{ij}\otimes\tilde{x}_{ij}\right)\right)}{Mag\left(\oplus_{i=1}^{m}\oplus_{j=1}^{n}\left(\tilde{d}_{ij}\otimes\tilde{x}_{ij}\right)\right)}\right)$

Subject to

Constraints of the problem (*P*5.6).

Step 3: Using the expression $Mag\left(\oplus_{i=1}^{k}\tilde{A}_i\right) = \sum_{i=1}^{k} Mag\left(\tilde{A}_i\right)$, discussed in Chap. 4, the crisp linear fractional programming problem (*P*5.7) can be transformed into its equivalent crisp linear fractional programming problem (*P*5.8).

Problem (*P*5.8)

Minimize $\left(Mag\left(\tilde{Z}\right) = \frac{\left(\sum_{i=1}^{m}\sum_{j=1}^{n} Mag\left(\tilde{c}_{ij}\otimes\tilde{x}_{ij}\right)\right)}{\left(\sum_{i=1}^{m}\sum_{j=1}^{n} Mag\left(\tilde{d}_{ij}\otimes\tilde{x}_{ij}\right)\right)}\right)$

Subject to

$$\sum_{j=1}^{n} Mag\left(\tilde{x}_{ij}\right) = or \geq or \leq Mag(\tilde{a}_i);\ i = 1, 2, \ldots, m;$$

$$\sum_{i=1}^{m} Mag\left(\tilde{x}_{ij}\right) = or \geq or \leq Mag\left(\tilde{b}_j\right);\ j = 1, 2, \ldots, n;$$

$$Mag\left(\tilde{x}_{ij}\right) \geq Mag\left(\tilde{0}\right);\ i = 1, 2, \ldots, m;\ j = 1, 2, \ldots, n.$$

Step 4: Using the expression $Mag\left(\otimes_{i=1}^{k}\tilde{A}_i\right) = \prod_{i=1}^{k} Mag\left(\tilde{A}_i\right)$, discussed in Chap. 4, the crisp linear fractional programming problem (*P*5.8) can be transformed into its equivalent crisp linear fractional programming problem ($P5.9$).

Problem ($P5.9$)

$$Minimize\left(Mag\left(\tilde{Z}\right)=\frac{\left(\sum_{i=1}^{m}\sum_{j=1}^{n}Mag(\tilde{c}_{ij})Mag(\tilde{x}_{ij})\right)}{\left(\sum_{i=1}^{m}\sum_{j=1}^{n}Mag(\tilde{d}_{ij})Mag(\tilde{x}_{ij})\right)}\right)$$

Subject to

Constraints of the problem ($P5.8$).

Step 5: Using the expression $Mag\left(\tilde{A}_i\right)=a_i^2$, discussed in Chap. 4, transform the crisp linear fractional programming problem ($P5.9$) into its equivalent crisp linear fractional programming problem ($P5.10$).

Problem ($P5.10$)

$$Minimize\left(Z^2=\frac{\sum_{i=1}^{m}\sum_{j=1}^{n}c_{ij}^2x_{ij}^2}{\sum_{i=1}^{m}\sum_{j=1}^{n}d_{ij}^2x_{ij}^2}\right)$$

Subject to

$$\sum_{j=1}^{n}x_{ij}^2 = or \geq or \leq a_i^2;\ i=1,2,\ldots,m$$

$$\sum_{i=1}^{m}x_{ij}^2 = or \geq or \leq b_j^2;\ j=1,2,\ldots,n$$

$$x_{ij}^2\geq 0;\ i=1,2,\ldots,m;\ j=1,2,\ldots,n.$$

Step 6: As discussed in Sect. 5.3, the crisp linear programming problem ($P5.5$) represents the dual of the crisp linear fractional programming problem ($P5.10$). Hence, the crisp linear programming problem ($P5.5$) represents the dual of the intuitionistic fuzzy linear fractional programming problem ($P5.4$).

5.5 Crisp Linear Fractional Transportation Problem with Mixed Constraints Equivalent to Symmetric Triangular Intuitionistic Fuzzy Linear Fractional Transportation Problem with Mixed Constraints

The following clearly indicates that the symmetric triangular intuitionistic fuzzy linear fractional transportation problem with mixed constraints, represented by Table 5.2, is equivalent to crisp linear fractional transportation problem with mixed constraints represented by Table 5.3.

It is obvious from Step 1 to Step 5 of Sect. 5.4 that the intuitionistic fuzzy linear fractional programming problem ($P5.4$) corresponding to the symmetric triangular intuitionistic fuzzy linear fractional transportation problem with mixed constraints, represented by Table 5.2, is equivalent to the crisp linear fractional programming problem ($P5.10$). Also, it is obvious from Sect. 5.3 that the crisp linear fractional

Table 5.3 Crisp linear fractional transportation problem with mixed constraints

	D_1	D_2	...	D_n	Availability
S_1	c_{11}^2 d_{11}^2	c_{12}^2 d_{12}^2	...	c_{1n}^2 d_{1n}^2	$=$ *or* $\leq$ *or* $\geq a_1^2$
S_2	c_{21}^2 d_{21}^2	c_{22}^2 d_{22}^2	...	c_{2n}^2 d_{2n}^2	$=$ *or* $\leq$ *or* $\geq a_2^2$
$\vdots$	$\vdots$	$\vdots$	$\vdots$	$\vdots$	$\vdots$
S_m	c_{m1}^2 d_{m1}^2	c_{m2}^2 d_{m2}^2	...	c_{mn}^2 d_{mn}^2	$=$ *or* $\leq$ *or* $\geq a_m^2$
Demand	$=$ *or* $\leq$ *or* $\geq b_1^2$	$=$ *or* $\leq$ *or* $\geq b_2^2$	...	$=$ *or* $\leq$ *or* $\geq b_n^2$	

programming problem ($P5.10$) is equivalent to crisp linear fractional transportation problem with mixed constraints represented by Table 5.3.

5.6 Proposed Sufficient Condition-IV for the Existence of at Least One More-For-Less Solution

It is obvious from Sect. 5.5 that the symmetric triangular intuitionistic fuzzy linear fractional transportation problem with mixed constraints, represented by Table 5.2, is equivalent to the crisp linear fractional transportation problem with mixed constraints represented by Table 5.3. Hence, to propose a sufficient condition-IV for the existence of at least one more-less-solution for the symmetric triangular intuitionistic fuzzy linear fractional transportation problem with mixed constraints, represented by Table 5.2, is equivalent to propose a sufficient condition-IV for the existence of at least one more-less-solution for the crisp linear fractional transportation problem with mixed constraints represented by Table 5.3. Keeping the same in mind, in this section, the following sufficient condition-IV for the existence of at least one more-for-less solution of crisp linear fractional transportation problems with mixed constraints, represented by Table 5.3, is proposed.

> If the condition (5.1a) and (5.1b) will be satisfied. Then, at least one more-for-less solution of the crisp linear fractional transportation problem with mixed constraints, represented by Table 5.3, will exist. Hence, at least one more-for-less solution of the symmetric triangular intuitionistic fuzzy linear fractional transportation problem with mixed constraints, represented by Table 5.2, will exist.

$$\sum_{i=1}^{m} a_i^2 u_i^2 + \sum_{j=1}^{n} b_j^2 v_j^2 + \sum_{i=1}^{m} \sum_{j=1}^{n} P_{ij} w_{ij}^2 = 0 \tag{5.1a}$$

$$u_p^2 + v_q^2 + Q_{pq} w_{pq}^2 < 0 \tag{5.1b}$$

where,

(i) $P_{ij} = \begin{cases} M; & \begin{array}{l} if\ on\ solving\ the\ problem\left(P5.1^{ij2}\right) \\ an\ unbounded\ optimal\ solution\ is\ obtained \end{array} \\ 0; & \begin{array}{l} if\ on\ solving\ the\ problem\left(P5.1^{ij2}\right) \\ a\ bounded\ optimal\ solution\ is\ obtained \end{array} \end{cases}$

(ii) $Q_{ij} = \begin{cases} 1; & \begin{array}{l} if\ on\ solving\ the\ problem\left(P5.1^{ij2}\right) \\ an\ unbounded\ optimal\ solution\ is\ obtained \end{array} \\ 0; & \begin{array}{l} if\ on\ solving\ the\ problem\left(P5.1^{ij2}\right) \\ a\ bounded\ optimal\ solution\ is\ obtained \end{array} \end{cases}$

(iii) $\left\{u_i^2, v_j^2, w_{ij}^2\right\}$ is an optimal solution of the crisp linear programming problem ($P5.5$).

5.6.1 Origin of the Sufficient Condition (5.1a)

The sufficient condition (5.1a) is obtained as follows.

It is pertinent to mention that as $\sum_{i=1}^{m} a_i^2 u_i^2 + \sum_{j=1}^{n} b_j^2 v_j^2 + \sum_{i=1}^{m} \sum_{j=1}^{n} P_{ij} w_{ij}^2$ represents the value of slack variable corresponding to the constraint $-\sum_{i=1}^{m} a_i^2 u_i^2 - \sum_{j=1}^{n} b_j^2 v_j^2 - \sum_{i=1}^{m} \sum_{j=1}^{n} P_{ij} w_{ij}^2 \le 0$ of the problem ($P5.5$). Therefore, if $\sum_{i=1}^{m} a_i^2 u_i^2 + \sum_{j=1}^{n} b_j^2 v_j^2 + \sum_{i=1}^{m} \sum_{j=1}^{n} P_{ij} w_{ij}^2 \neq 0$ i.e., the value of slack variable is not equal to zero. Then, by complementary slackness theorem [5], the value of dual variable $t = \frac{1}{\sum_{i \in I} \sum_{j \in J} d_{ij}^2 x_{ij}^2}$ in the optimal solution of the problem ($P5.11$) will be 0. This indicates that at least one x_{ij} tends to $+\infty$ in the optimal solution of the problem ($P5.10'$) i.e., the total optimal quantity of the supplied product tends to $+\infty$. Therefore, it is not possible to increase the total optimal quantity of the supplied product. Hence, no more-for-less solution will exist if $\sum_{i=1}^{m} a_i^2 u_i^2 + \sum_{j=1}^{n} b_j^2 v_j^2 + \sum_{i=1}^{m} \sum_{j=1}^{n} P_{ij} w_{ij}^2 \neq 0$.

Problem ($P5.11$) *Minimize* $\left(\sum_{i=1}^{m} \sum_{j=1}^{n} c_{ij}^2 y_{ij}^2\right)$

Subject to

$$\sum_{i=1}^{m} \sum_{j=1}^{n} d_{ij}^2 y_{ij}^2 = 1,$$

$$\sum_{j=1}^{n} y_{ij}^2 - a_i^2 t = or \ge or \le 0;\ i = 1, 2, \ldots, m;$$

$$\sum_{i=1}^{n} y_{ij}^2 - b_j^2 t = or \geq or \leq 0;\ j = 1, 2, \ldots, n;$$

$$y_{ij}^2 - Mt \leq 0,$$

$$y_{ij}^2 \geq 0;\ i = 1, 2, \ldots, m;\ j = 1, 2, \ldots, n.$$

$$t \geq 0.$$

Problem $(P5.10')$ *Minimize* $\left(Z^2 = \frac{\sum_{i=1}^{m}\sum_{j=1}^{n} c_{ij}^2 x_{ij}^2}{\sum_{i=1}^{m}\sum_{j=1}^{n} d_{ij}^2 x_{ij}^2}\right)$
Subject to
Constraints of the problem ($P5.10$) with the additional constraints
$x_{pq}^2 \leq M$ for all those p, q corresponding to which there exists an unbounded optimal solution of the problem ($P5.10^{pq2}$), where M is a very large positive real number tends to $+\infty$.

Problem $(P5.10^{pq2})$ *Maximize* (x_{pq}^2)
Subject to
Constraints of the problem ($P5.10$).

5.6.2 Origin of the Sufficient Condition (5.1b)

The sufficient condition (5.1b) is obtained as follows.

If the availability of the pth source and the demand of the qth destination is increased by a positive quantity (say, θ) in the dual problem ($P5.5$). Then, the modified dual problem ($P5.12$) will be obtained.

Problem (***P*5.12**) *Maximize* (k)
Subject to

$$-\sum_{i=1}^{m} a_i^2 u_i^2 - \sum_{j=1}^{n} b_j^2 v_j^2 - \sum_{i=1}^{m}\sum_{j=1}^{n} P_{ij} w_{ij}^2 - \theta\left(u_p^2 + v_q^2\right) \leq 0,$$

$$u_i^2 + v_j^2 + Q_{ij} w_{ij}^2 + d_{ij}^2 k \leq c_{ij}^2;\ i = 1, 2, \ldots, m;\ j = 1, 2, \ldots, n,$$

$$u_i^2 * 0,\ v_j^2 \circ 0;\ i = 1, 2, \ldots, m;\ j = 1, 2, \ldots, n,$$

$$w_{ij}^2 \leq 0,$$

k. is unrestricted dual variable,
where,

$$\text{(i)} \quad P_{ij} = \begin{cases} M; & \begin{array}{l} if\ on\ solving\ the\ problem\big(P5.1^{ij2}\big) \\ an\ unbounded\ optimal\ solution\ is\ obtained \end{array} \\ 0; & \begin{array}{l} if\ on\ solving\ the\ problem\big(P5.1^{ij2}\big) \\ a\ bounded\ optimal\ solution\ is\ obtained \end{array} \end{cases}$$

$$\text{(ii)} \quad Q_{ij} = \begin{cases} 1; & \begin{array}{l} if\ on\ solving\ the\ problem\big(P5.1^{ij2}\big) \\ an\ unbounded\ optimal\ solution\ is\ obtained \end{array} \\ 0; & \begin{array}{l} if\ on\ solving\ the\ problem\big(P5.1^{ij2}\big) \\ a\ bounded\ optimal\ solution\ is\ obtained \end{array} \end{cases}$$

It is obvious from the constraint

$$u_i^2 + v_j^2 + Q_{ij} w_{ij}^2 + d_{ij}^2 k \le c_{ij}^2$$

of the dual problem $(P5.12)$ that

$$k \le \frac{c_{ij}^2 - \left(u_i^2 + v_j^2 + Q_{ij} w_{ij}^2\right)}{d_{ij}^2}.$$

Since, the above condition is valid for all i, j. So, it will also be valid for $i = p$ and $j = q$ i.e.,

$$k \le \frac{c_{pq}^2 - \left(u_p^2 + v_q^2 + Q_{pq} w_{pq}^2\right)}{d_{pq}^2}.$$

It is pertinent to mention that to decrease the optimal value of the crisp linear fractional programming problem (P5.10′) is equivalent to increase the optimal value of the dual problem ($P5.12$). Also, it is obvious from ($P5.12$) that the optimal value of the dual problem ($P5.12$) i.e., k will increase if $u_p^2 + v_q^2 + Q_{pq} w_{pq}^2 < 0$.

5.7 Proposed Mehar Method-IV

In this section, a new method (named as Mehar method-IV) is proposed to check the existence of at least one more-for-less solution as well as to find all more-for-less solutions (if exists) of symmetric triangular intuitionistic fuzzy linear fractional transportation problems with mixed constraints.

The steps of the proposed Mehar method-IV are as follows.

Step 1: Use **Step 1.1** to **Step 1.5** to check that for the symmetric triangular intuitionistic fuzzy linear fractional transportation problem with mixed constraints, represented by Table 5.2, at least one more-for-less solution will exist or not.

Step 1.1: Transform the symmetric triangular intuitionistic fuzzy linear fractional transportation problem with mixed constraints, represented by Table 5.2, into its equivalent crisp linear fractional transportation problem with mixed constraints represented by Table 5.3.

Step 1.2: Find an optimal solution $\left\{u_i^2, v_j^2, w_{ij}^2\right\}$ of the problem ($P5.5$) by substituting the values of $c_{ij}^2, d_{ij}^2, a_i^2$ and b_j^2 from Table 5.3.

Step 1.3: Using the values of u_i^2, v_j^2, w_{ij}^2, obtained in **Step 1.2**, check if the condition (5.1a) is satisfying or not.

Case (i): If the condition (5.1a) is satisfying i.e., if $\sum_{i=1}^{m} a_i^2 u_i^2 + \sum_{j=1}^{n} b_j^2 v_j^2 + \sum_{i=1}^{m}\sum_{j=1}^{n} P_{ij} w_{ij}^2 = 0$. Then, go to **Step 1.4**.

Case (ii): If the condition (5.1a) is not satisfying i.e., if $\sum_{i=1}^{m} a_i^2 u_i^2 + \sum_{j=1}^{n} b_j^2 v_j^2 + \sum_{i=1}^{m}\sum_{j=1}^{n} P_{ij} w_{ij}^2 \neq 0$. Then, there does not exist a more-for-less solution as it is not possible to increase the optimal supplied quantity of the product.

Step 1.4: Using the values of u_i^2, v_j^2, w_{ij}^2, obtained in **Step 1.2**, construct a shadow price matrix $SP = \left[u_i^2 + v_j^2 + L_{ij} w_{ij}^2\right]_{m \times n}$, where

(i) $L_{ij} = 0$; corresponding to those i, j for which a bounded optimal solution of the problem ($P5.1^{ij2}$) is obtained.

(ii) $L_{ij} = 1$; corresponding to those i, j for which there exists an unbounded optimal solution of the problem ($P5.1^{ij2}$).

Step 1.5: Check that all the elements of the shadow price matrix, obtained in **Step 1.4**, are non-negative or not.

Case (i): If all the elements of the shadow price matrix, obtained in **Step 1.4**, are non-negative. Then, there does not exist a more-for-less solution.

Case (ii): If at least one element of the shadow price matrix, obtained in **Step 1.4**, is negative. Then, at least one more-for-less solution will exist.

Step 2: If the element corresponding to the pth row and the qth column of the shadow price matrix, obtained in **Step 1.4**, is negative. Then, a more-for-less solution can be obtained by increasing the availability of the pth source and the demand of the qth destination with the help of **Step 2.1** to **Step 2.10** of the post-optimality analysis [5].

Step 2.1: Construct the matrix R_{pq}

$$R_{pq} = \begin{bmatrix} a_1^2 \\ a_2^2 \\ \vdots \\ a_p^2 + \theta \\ \vdots \\ a_m^2 \\ b_1^2 \\ b_2^2 \\ \vdots \\ b_q^2 + \theta \\ \vdots \\ b_n^2 \\ \infty \\ \vdots \\ \infty \end{bmatrix}$$

where,

(i) θ is a positive real number.

(ii) Number of ∞ in the matrix R_{pq} will be equal to the number of additional constraints in the modified crisp linear fractional programming problem ($P5.10'$).

Step 2.2: Using **Step 2.2.1** and **Step 2.2.2** or any other appropriate method, find an optimal solution $\left\{x_{ij}^2; i = 1, 2, \ldots, m; j = 1, 2, \ldots, n\right\}$ of the modified crisp linear fractional programming problem ($P5.10'$).

Step 2.2.1: Find an optimal solution $\left\{t, y_{ij}^2; i = 1, 2, \ldots, m; j = 1, 2, \ldots, n\right\}$ of the crisp linear programming problem ($P5.11$).

Step 2.2.2: Using the optimal solution, obtained in **Step 2.2.1**, find an optimal solution $\left\{x_{ij}^2 = \lim\limits_{M \to \infty}\left(\frac{y_{ij}^2}{t}\right); i = 1, 2, \ldots, m; j = 1, 2, \ldots, n\right\}$ of the modified crisp linear fractional programming problem ($P5.10'$).

Step 2.3: Substituting the optimal values of decision variables x_{ij}^2, obtained in **Step 2.2.2**, find the optimal value of slack and/or surplus variable corresponding to each constraint of the modified crisp linear fractional programming problem ($P5.10'$).

Step 2.4: Identify which optimal variables (decision variables, slack variables, surplus variables), obtained in **Step 2.2.2** and **Step 2.3**, are basic variables.

Step 2.5: Construct a basis matrix B with the help of the coefficients of the identified optimal basic variables in the constraints of the modified crisp linear fractional programming problem ($P5.10'$).

Step 2.6: Find the multiplicative inverse i.e., B^{-1} of the basis matrix B.
Step 2.7: Find the matrix $B^{-1}R_{pq}$.
Step 2.8: Find the range of θ(where, $\theta > 0$) for which all the elements of the matrix $B^{-1}R_{pq}$, obtained in **Step 2.7**, are greater than or equal to zero.
Step 2.9: For any value of θ lying in its range, obtained in **Step 2.8**, the ith element (say, τ_i) of the matrix $B^{-1}R_{pq}$ represents the value of the ith basic variable in a more-for-less solution of crisp linear fractional transportation problem with mixed constraints represented by Table 5.3.
Step 2.10: For any value of θ lying in its range, obtained in **Step 2.8**, the symmetric triangular intuitionistic fuzzy number $(\tau_i, \delta_{i1}(1-\alpha), \delta_{i1}(1-\alpha); \tau_i, \delta_{i2}(1-\alpha), \delta_{i2}(1-\alpha))$, where δ_{i1} and δ_{i2} are any non-negative real numbers and $\alpha \in (0, 1]$, represents the value of the ith basic intuitionistic fuzzy variable in a more-for-less solution of the symmetric triangular intuitionistic fuzzy linear fractional transportation problem with mixed constraints represented by Table 5.2.

5.8 Illustrative Examples

In this section, to illustrate the proposed Mehar method-IV, all more-for-less solutions of two existing crisp linear fractional transportation problems with mixed constraints [4] as well as a symmetric triangular intuitionistic fuzzy linear fractional transportation problem with mixed constraints are obtained by the proposed Mehar method-IV.

5.8.1 *All More-For-Less Solutions of the First Existing Problem*

Gupta et al. [4] considered the crisp linear fractional transportation problem with mixed constraints, represented by Table 5.4, to illustrate their proposed method. In this section, all more-for-less solutions (if exists) of the same problem are obtained by the proposed Mehar method-IV.

Using the proposed Mehar method-IV, all more-for-less solutions (if exists) of the crisp linear fractional transportation problem with mixed constraints, represented by Table 4.4, can be obtained as follows.

Step 1: According to **Step 1** of the proposed Mehar method-IV, there is a need to use **Step 1.1** to **Step 1.5** to check that for the crisp transportation problem with mixed constraints, represented by Table 5.4, at least one more-for-less solution will exist or not.

Step 1.1: According to **Step 1.1** of the proposed Mehar method-IV, there is a need to transform the symmetric triangular intuitionistic fuzzy linear fractional

Table 5.4 First crisp linear fractional transportation problem with mixed constraints [4]

	D_1	D_2	D_3	Availability
S_1	$c_{11}^2 = 5$ $d_{11}^2 = 6$	$c_{12}^2 = 4$ $d_{12}^2 = 3$	$c_{13}^2 = 2$ $d_{13}^2 = 4$	$= 5$
S_2	$c_{21}^2 = 6$ $d_{21}^2 = 7$	$c_{22}^2 = 5$ $d_{22}^2 = 4$	$c_{23}^2 = 3$ $d_{23}^2 = 2$	≥ 10
S_3	$c_{31}^2 = 8$ $d_{31}^2 = 6$	$c_{32}^2 = 9$ $d_{32}^2 = 5$	$c_{33}^2 = 4$ $d_{33}^2 = 2$	≤ 9
Demand	$= 8$	≥ 15	≤ 6	

transportation problem with mixed constraints into its equivalent crisp linear fractional transportation problem with mixed constraints. Since, the linear fractional transportation problem with mixed constraints, represented by Table 5.4, is a crisp linear fractional transportation problem with mixed constraints. So, there is no need to apply this step.

Step 1.2: According to **Step 1.2** of the proposed Mehar method-IV, there is a need to find an optimal solution of the crisp linear programming problem ($P5.13$) which is obtained by substituting $P_{22} = M$, $Q_{22} = 1$, $P_{11} = P_{12} = P_{13} = P_{21} = P_{23} = P_{31} = P_{32} = P_{33} = Q_{11} = Q_{12} = Q_{13} = Q_{21} = Q_{23} = Q_{31} = Q_{32} = Q_{33} = 0$ as on solving the problem ($P5.14^{222}$), an unbounded optimal solution is obtained. While, on solving the problems ($P5.14^{112}$), ($P5.14^{112}$), ($P5.14^{132}$), ($P5.14^{212}$), ($P5.14^{232}$), ($P5.14^{312}$), ($P5.14^{322}$) and ($P5.14^{332}$), a bounded optimal solution is obtained.

Problem (*P*5.13)

Maximize (k)

Subject to

$$-5u_1^2 - 10u_2^2 - 9u_3^2 - 8v_1^2 - 15v_2^2 - 6v_3^2 - Mw_{22}^2 \leq 0,$$

$$u_1^2 + v_1^2 + 6k \leq 5,$$

$$u_1^2 + v_2^2 + 3k \leq 4,$$

$$u_1^2 + v_3^2 + 4k \leq 2,$$

$$u_2^2 + v_1^2 + 7k \leq 6,$$

$$u_2^2 + v_2^2 + w_{22}^2 + 4k \leq 5,$$

$$u_2^2 + v_3^2 + 2k \leq 3,$$

$$u_3^2 + v_1^2 + 6k \leq 8,$$

$$u_3^2 + v_2^2 + 5k \leq 9,$$

$$u_3^2 + v_3^2 + 2k \leq 4,$$

$$u_2^2, v_2^2 \geq 0,$$

$$u_3^2, v_3^2, w_{22}^2 \leq 0,$$

u_1^2, v_1^2, k are unrestricted in sign.

Problem $\left(P5.14^{112}\right)$
Maximize $\left(x_{11}^2\right)$
Subject to

$$x_{11}^2 + x_{12}^2 + x_{13}^2 = 5,$$

$$x_{21}^2 + x_{22}^2 + x_{23}^2 \geq 10,$$

$$x_{31}^2 + x_{32}^2 + x_{33}^2 \leq 9,$$

$$x_{11}^2 + x_{21}^2 + x_{31}^2 = 8,$$

$$x_{12}^2 + x_{22}^2 + x_{32}^2 \geq 15,$$

$$x_{13}^2 + x_{23}^2 + x_{33}^2 \leq 6,$$

$$x_{ij} \geq 0, i = 1, 2, 3; \; j = 1, 2, 3.$$

Problem $\left(P5.14^{122}\right)$
Maximize $\left(x_{12}^2\right)$
Subject to
Constraints of the problem $(P5.14^{112})$.

Problem $\left(P5.14^{132}\right)$
Maximize $\left(x_{13}^2\right)$
Subject to
Constraints of the problem $(P5.14^{112})$.

Problem $(P5.14^{212})$
Maximize (x_{21}^2)
Subject to
Constraints of the problem $(P5.14^{112})$.
Problem $(P5.14^{222})$
Maximize (x_{22}^2)
Subject to
Constraints of the problem $(P5.14^{112})$.
Problem $(P5.14^{232})$
Maximize (x_{23}^2)
Subject to
Constraints of the problem $(P5.14^{112})$.
Problem $(P5.14^{312})$
Maximize (x_{31}^2)
Subject to
Constraints of the problem $(P5.14^{112})$.
Problem $(P5.14^{322})$
Maximize (x_{32}^2)
Subject to
Constraints of the problem $(P5.14^{112})$.
Problem $(P5.14^{332})$
Maximize (x_{33}^2)
Subject to
Constraints of the problem $(P5.14^{112})$.

It can be easily verified that on solving the problem $(P5.13)$, the obtained optimal solution is $u_1^2 = -\frac{65}{34}$, $u_2^2 = 0$, $u_3^2 = 0$, $v_1^2 = -\frac{115}{136}$, $v_2^2 = \frac{37}{34}$, $v_3^2 = 0$, $w_{22}^2 = 0$.

Step 1.3: It is obvious that $5u_1^2 + 10u_2^2 + 9u_3^2 + 8v_1^2 + 15v_2^2 + 6v_3^2 + Mw_{22}^2 = 0$. So, according to Case (i) of **Step 1.3** of the proposed Mehar method-IV, there is a need to go to **Step 1.4**.

Step 1.4: According to **Step 1.4** of the proposed Mehar method-IV, the following shadow price matrix is obtained.

$$SP = \begin{bmatrix} u_1^2 + v_1^2 = -\frac{375}{136} & u_1^2 + v_2^2 = -\frac{14}{17} & u_1^2 + v_3^2 = -\frac{65}{34} \\ u_2^2 + v_1^2 = -\frac{115}{136} & u_2^2 + v_2^2 + w_{22}^2 = \frac{37}{34} & u_2^2 + v_3^2 = 0 \\ u_3^2 + v_1^2 = -\frac{115}{136} & u_3^2 + v_2^2 = \frac{37}{34} & u_3^2 + v_3^2 = 0 \end{bmatrix}$$

Step 1.5: Since, some elements of the shadow price matrix, obtained in **Step 1.4**, are negative. So, according to Case (ii) of **Step 1.5** of the proposed Mehar method-IV, there exists at least one more-for-less solution of the crisp transportation problem with mixed constraints, represented by Table 5.4.

Step 2: According to Step 2 of the proposed Mehar method-IV, all more-for-less solutions of the crisp transportation problem with mixed constraints, represented by Table 5.4, can be obtained as follows.

Step 2.1: It is obvious from **Step 1.4** that $u_1^2+v_1^2<0$, $u_1^2+v_2^2<0$, $u_1^2+v_3^2<0$, $u_2^2+v_1^2<0$ and $u_3^2+v_1^2<0$. Therefore, according to **Step 2.1** of the proposed Mehar method-IV,

i. $R_{11}=\begin{bmatrix}5+\theta\\10\\9\\8+\theta\\15\\6\\\infty\end{bmatrix}$

ii. $R_{12}=\begin{bmatrix}5+\theta\\10\\9\\8\\15+\theta\\6\\\infty\end{bmatrix}$

iii. $R_{13}=\begin{bmatrix}5+\theta\\10\\9\\8\\15\\6+\theta\\\infty\end{bmatrix}$

iv. $R_{21}=\begin{bmatrix}5\\10+\theta\\9\\8+\theta\\15\\6\\\infty\end{bmatrix}$

v. $R_{31}=\begin{bmatrix}5\\10\\9+\theta\\8+\theta\\15\\6\\\infty\end{bmatrix}$

It is pertinent to mention that the number of ∞ in the matrices R_{11}, R_{12}, R_{13}, R_{21} and R_{31} will be equal to 1 as the number of additional constraints in the modified crisp linear fractional programming problem ($P5.14'$) is 1.

Problem $(\boldsymbol{P}5.14')$

Minimize $\left(\frac{5x_{11}^2+4x_{12}^2+2x_{13}^2+6x_{21}^2+5x_{22}^2+3x_{23}^2+8x_{31}^2+9x_{32}^2+4x_{33}^2}{6x_{11}^2+3x_{12}^2+4x_{13}^2+7x_{21}^2+4x_{22}^2+2x_{23}^2+6x_{31}^2+5x_{32}^2+2x_{33}^2}\right)$

Subject to

Constraints of the problem ($P5.14^{112}$) with the additional constraint $x_{22}^2 \leq M$

Step 2.2: According to the proposed Mehar method-IV, there is a need to find an optimal solution of the modified crisp linear fractional programming problem ($P5.14'$).

Step 2.2.1 According to the proposed Mehar method-IV, there is a need to find an optimal solution of the crisp linear programming problem ($P5.15$).

Problem $(\boldsymbol{P}5.15)$

Minimize $\left(5y_{11}^2+4y_{12}^2+2y_{13}^2+6y_{21}^2+5y_{22}^2+3y_{23}^2+8y_{31}^2+9y_{32}^2+4y_{33}^2\right)$

Subject to

$$6y_{11}^2+3y_{12}^2+4y_{13}^2+7y_{21}^2+4y_{22}^2+2y_{23}^2+6y_{31}^2+5y_{32}^2+2y_{33}^2=1,$$

$$y_{11}^2+y_{12}^2+y_{13}^2=5t,$$

$$y_{21}^2+y_{22}^2+y_{23}^2\geq 10t,$$

$$y_{31}^2+y_{32}^2+y_{33}^2\leq 9t,$$

$$y_{11}^2+y_{21}^2+y_{31}^2=8t,$$

$$y_{12}^2+y_{22}^2+y_{32}^2\geq 15t,$$

$$y_{13}^2+y_{23}^2+y_{33}^2\leq 6t,$$

$$y_{22}^2\leq Mt,$$

$$y_{ij}^2\geq 0, i=1,2,3;\ j=1,2,3.$$

It can be easily verified that the following optimal solution is obtained on solving the crisp linear programming problem (P5.15).

$$y_{11}^2=0, y_{12}^2=0, y_{13}^2=\frac{5}{136}, y_{21}^2=\frac{8}{136}, y_{22}^2=\frac{15}{136}, y_{23}^2=0, y_{31}^2=0, y_{32}^2=0, y_{33}^2=0, t=\frac{1}{136}.$$

Step 2.2.2: According to the proposed Mehar method-IV, $\{x_{11}^2 = \lim_{M\to\infty} \frac{y_{11}^2}{t} = 0, x_{12}^2 = \lim_{M\to\infty} \frac{y_{12}^2}{t} = 0, x_{13}^2 = \lim_{M\to\infty} \frac{y_{13}^2}{t} = \lim_{M\to\infty} \frac{\frac{5}{136}}{\frac{1}{136}} = 5, x_{21}^2 = \lim_{M\to\infty} \frac{y_{21}^2}{t} = \lim_{M\to\infty} \frac{\frac{8}{136}}{\frac{1}{136}} = 8, x_{22}^2 = \lim_{M\to\infty} \frac{y_{22}^2}{t} = \lim_{M\to\infty} \frac{\frac{15}{136}}{\frac{1}{136}} = 15, x_{23}^2 = \lim_{M\to\infty} \frac{y_{23}^2}{t} = 0, x_{31}^2 = \lim_{M\to\infty} \frac{y_{31}^2}{t} = 0, x_{32}^2 = \lim_{M\to\infty} \frac{y_{32}^2}{t} = 0, x_{33}^2 = \lim_{M\to\infty} \frac{y_{33}^2}{t} = 0\}$represents an optimal solution of the modified crisp linear fractional programming problem ($P5.14'$) and the corresponding total optimal transportation cost is $\frac{5x_{11}^2+4x_{12}^2+2x_{13}^2+6x_{21}^2+5x_{22}^2+3x_{23}^2+8x_{31}^2+9x_{32}^2+4x_{33}^2}{6x_{11}^2+3x_{12}^2+4x_{13}^2+7x_{21}^2+4x_{22}^2+2x_{23}^2+6x_{31}^2+5x_{32}^2+2x_{33}^2} = \frac{5(0)+4(0)+2(5)+6(8)+5(15)+3(0)+8(0)+9(0)+4(0)}{6(0)+3(0)+4(5)+7(8)+4(15)+2(0)+6(0)+5(0)+2(0)} = \frac{133}{136} = 0.9779.$

Step 2.3: According to **Step 2.3** of the proposed Mehar method-IV, there is a need to find the optimal values of slack and/or surplus variables corresponding to each constraint of the modified crisp linear fractional programming problem ($P5.14'$).

(i) Since, in the first row constraint i.e., $x_{11}^2 + x_{12}^2 + x_{13}^2 = 5$, of the modified crisp linear fractional programming problem ($P5.14'$), the sign is equal sign. So, neither any slack variable nor any surplus variable exists corresponding to this constraint.

(ii) Putting $x_{21}^2 = 8, x_{22}^2 = 15, x_{23}^2 = 0$ in the second row constraint i.e., $x_{21}^2 + x_{22}^2 + x_{23}^2 \geq 10$, of the modified crisp linear fractional programming problem ($P5.14'$), the optimal value of the surplus variable (say, s_1^2) i.e., $s_1^2 = x_{21}^2 + x_{22}^2 + x_{23}^2 - 10$ is 13.

(iii) Putting $x_{31}^2 = 0, x_{32}^2 = 0, x_{33}^2 = 0$ in the third row constraint i.e., $x_{31}^2 + x_{32}^2 + x_{33}^2 \leq 9$, of the modified crisp linear fractional programming problem ($P5.14'$), the optimal value of the slack variable (say, s_2^2) i.e., $s_2^2 = 9 - x_{31}^2 - x_{32}^2 - x_{33}^2$ is 9.

(iv) Since, in the first column constraint i.e., $x_{11}^2 + x_{21}^2 + x_{31}^2 = 8$, of the modified crisp linear fractional programming problem ($P5.14'$), the sign is equal sign. So, neither any slack variable nor any surplus variable exists corresponding to this constraint.

(v) Putting $x_{12}^2 = 0, x_{22}^2 = 15, x_{32}^2 = 0$ in the second column constraint i.e., $x_{12}^2 + x_{22}^2 + x_{32}^2 \geq 15$, of the modified crisp linear fractional programming problem ($P5.14'$), the optimal value of the surplus variable (say, s_3^2) i.e., $s_3^2 = x_{12}^2 + x_{22}^2 + x_{32}^2 - 15$ is 0.

(vi) Putting $x_{13}^2 = 5, x_{23}^2 = 0, x_{33}^2 = 0$ in the third column constraint i.e., $x_{13}^2 + x_{23}^2 + x_{33}^2 \leq 6$, of the modified crisp linear fractional programming problem ($P5.14'$), the optimal value of the slack variable (say, s_4^2) i.e., $s_4^2 = 6 - x_{13}^2 - x_{23}^2 - x_{33}^2$ is 1.

(vii) Putting $x_{22}^2 = 15$ in the additional constraint i.e., $x_{22}^2 \leq M$, of the modified crisp linear fractional programming problem ($P5.14'$), the optimal value of the slack variable (say, s_{22}^2) i.e., $s_{22}^2 = \lim_{M\to\infty}(M - 15)$ is ∞.

Step 2.4: According to **Step 2.4** of the proposed Mehar method-IV, there is a need to identify which optimal variables (decision variables, slack variables, surplus variables), obtained in **Step 2.2.2** and **Step 2.3**, are basic variables.

It can be easily verified from **Step 2.2.2** and **Step 2.3** that the basic variables are $x_{13}^2, x_{21}^2, x_{22}^2, s_1^2, s_2^2, s_4^2$ and s_{22}^2

Step 2.5: According to **Step 2.5** of the proposed Mehar method-IV, there is a need to construct the basis matrix B with the help of the coefficients of optimal basic variables $x_{13}^2, x_{21}^2, x_{22}^2, s_1^2, s_2^2, s_4^2, s_{22}^2$ in the constraints of the modified crisp linear fractional programming problem ($P5.14'$).

$$B = \begin{bmatrix} 1 & 0 & 0 & 0 & 0 & 0 & 0 \\ 0 & 1 & 1 & -1 & 0 & 0 & 0 \\ 0 & 0 & 0 & 0 & 1 & 0 & 0 \\ 0 & 1 & 0 & 0 & 0 & 0 & 0 \\ 0 & 0 & 1 & 0 & 0 & 0 & 0 \\ 1 & 0 & 0 & 0 & 0 & 1 & 0 \\ 0 & 0 & 1 & 0 & 0 & 0 & 1 \end{bmatrix}$$

Step 2.6: According to **Step 2.6** of the proposed Mehar method-IV, there is a need to find the multiplicative inverse of the basis matrix B.

$$B^{-1} = \begin{bmatrix} 1 & 0 & 0 & 0 & 0 & 0 & 0 \\ 0 & 0 & 0 & 1 & 0 & 0 & 0 \\ 0 & 0 & 0 & 0 & 1 & 0 & 0 \\ 0 & -1 & 0 & 1 & 1 & 0 & 0 \\ 0 & 0 & 1 & 0 & 0 & 0 & 0 \\ -1 & 0 & 0 & 0 & 0 & 1 & 0 \\ 0 & 0 & 0 & 0 & -1 & 0 & 1 \end{bmatrix}$$

Step 2.7: According to **Step 2.7** of the proposed Mehar method-IV, there is a need to find $B^{-1}R_{11}$, $B^{-1}R_{12}$, $B^{-1}R_{13}$, $B^{-1}R_{21}$ and $B^{-1}R_{31}$.

(i) $$B^{-1}R_{11} = \begin{bmatrix} 1 & 0 & 0 & 0 & 0 & 0 & 0 \\ 0 & 0 & 0 & 1 & 0 & 0 & 0 \\ 0 & 0 & 0 & 0 & 1 & 0 & 0 \\ 0 & -1 & 0 & 1 & 1 & 0 & 0 \\ 0 & 0 & 1 & 0 & 0 & 0 & 0 \\ -1 & 0 & 0 & 0 & 0 & 1 & 0 \\ 0 & 0 & 0 & 0 & -1 & 0 & 1 \end{bmatrix} \begin{bmatrix} 5+\theta \\ 10 \\ 9 \\ 8+\theta \\ 15 \\ 6 \\ \infty \end{bmatrix} = \begin{bmatrix} 5+\theta \\ 8+\theta \\ 15+0\theta \\ 13+\theta \\ 9+0\theta \\ 1-\theta \\ \infty \end{bmatrix}$$

(ii) $$B^{-1}R_{12} = \begin{bmatrix} 1 & 0 & 0 & 0 & 0 & 0 & 0 \\ 0 & 0 & 0 & 1 & 0 & 0 & 0 \\ 0 & 0 & 0 & 0 & 1 & 0 & 0 \\ 0 & -1 & 0 & 1 & 1 & 0 & 0 \\ 0 & 0 & 1 & 0 & 0 & 0 & 0 \\ -1 & 0 & 0 & 0 & 0 & 1 & 0 \\ 0 & 0 & 0 & 0 & -1 & 0 & 1 \end{bmatrix} \begin{bmatrix} 5+\theta \\ 10 \\ 9 \\ 8 \\ 15+\theta \\ 6 \\ \infty \end{bmatrix} = \begin{bmatrix} 5+\theta \\ 8+0\theta \\ 15+\theta \\ 13+\theta \\ 9+0\theta \\ 1-\theta \\ \infty \end{bmatrix}$$

(iii) $$B^{-1}R_{13} = \begin{bmatrix} 1 & 0 & 0 & 0 & 0 & 0 & 0 \\ 0 & 0 & 0 & 1 & 0 & 0 & 0 \\ 0 & 0 & 0 & 0 & 1 & 0 & 0 \\ 0 & -1 & 0 & 1 & 1 & 0 & 0 \\ 0 & 0 & 1 & 0 & 0 & 0 & 0 \\ -1 & 0 & 0 & 0 & 0 & 1 & 0 \\ 0 & 0 & 0 & 0 & -1 & 0 & 1 \end{bmatrix} \begin{bmatrix} 5+\theta \\ 10 \\ 9 \\ 8 \\ 15 \\ 6+\theta \\ \infty \end{bmatrix} = \begin{bmatrix} 5+\theta \\ 8+0\theta \\ 15+0\theta \\ 13+0\theta \\ 9+0\theta \\ 1+0\theta \\ \infty \end{bmatrix}$$

(iv) $$B^{-1}R_{21} = \begin{bmatrix} 1 & 0 & 0 & 0 & 0 & 0 & 0 \\ 0 & 0 & 0 & 1 & 0 & 0 & 0 \\ 0 & 0 & 0 & 0 & 1 & 0 & 0 \\ 0 & -1 & 0 & 1 & 1 & 0 & 0 \\ 0 & 0 & 1 & 0 & 0 & 0 & 0 \\ -1 & 0 & 0 & 0 & 0 & 1 & 0 \\ 0 & 0 & 0 & 0 & -1 & 0 & 1 \end{bmatrix} \begin{bmatrix} 5 \\ 10+\theta \\ 9 \\ 8+\theta \\ 15 \\ 6 \\ \infty \end{bmatrix} = \begin{bmatrix} 5+0\theta \\ 8+\theta \\ 15+0\theta \\ 13+0\theta \\ 9+0\theta \\ 1+0\theta \\ \infty \end{bmatrix}$$

(v) $$B^{-1}R_{31} = \begin{bmatrix} 1 & 0 & 0 & 0 & 0 & 0 & 0 \\ 0 & 0 & 0 & 1 & 0 & 0 & 0 \\ 0 & 0 & 0 & 0 & 1 & 0 & 0 \\ 0 & -1 & 0 & 1 & 1 & 0 & 0 \\ 0 & 0 & 1 & 0 & 0 & 0 & 0 \\ -1 & 0 & 0 & 0 & 0 & 1 & 0 \\ 0 & 0 & 0 & 0 & -1 & 0 & 1 \end{bmatrix} \begin{bmatrix} 5 \\ 10 \\ 9+\theta \\ 8+\theta \\ 15 \\ 6 \\ \infty \end{bmatrix} = \begin{bmatrix} 5+0\theta \\ 8+\theta \\ 15+0\theta \\ 13+\theta \\ 9+\theta \\ 1+0\theta \\ \infty \end{bmatrix}$$

Step 2.8: According to **Step 2.8** of the proposed Mehar method-IV, there is a need to find the range of θ (where, $\theta > 0$) so that all the elements of the matrix, obtained in **Step 2.7**, are greater than or equal to zero.

It is obvious from **Step 2.7** that

(i) The range of θ for which all the elements of the matrix $B^{-1}R_{11}$, obtained in **Step 2.7**, are greater than or equal to zero, i.e., $5+\theta \geq 0, 8+\theta \geq 0, 13+\theta \geq 0, 1-\theta \geq 0$ is $0 < \theta \leq 1$.

(ii) The range of θ for which all the elements of the matrix $B^{-1}R_{12}$, obtained in **Step 2.7**, are greater than or equal to zero, i.e., $5+\theta \geq 0, 15+\theta \geq 0, 13+\theta \geq 0, 1-\theta \geq 0$ is $0 < \theta \leq 1$.

(iii) The range of θ for which all the elements of the matrix $B^{-1}R_{13}$, obtained in **Step 2.7**, are greater than or equal to zero, i.e., $5+\theta \geq 0$, is $\theta > 0$.

(iv) The range of θ for which all the elements of the matrix $B^{-1}R_{21}$, obtained in **Step 2.7**, are greater than or equal to zero, i.e., $8+\theta \geq 0$, is $\theta > 0$.

(v) The range of θ for which all the elements of the matrix $B^{-1}R_{31}$, obtained in **Step 2.7**, are greater than or equal to zero, i.e., $8+\theta \geq 0, 13+\theta \geq 0, 9+\theta \geq 0$ is $\theta > 0$.

Step 2.9: According to **Step 2.9** of the proposed Mehar method-IV,

(i) For any value θ such that $0 < \theta \leq 1$,
$5+\theta$ represents the value of the 1st basic variable i.e., x_{13}^2,
$8+\theta$ represents the value of the 2nd basic variable i.e., x_{21}^2,
$15+0\theta$ represents the value of the 3rd variable i.e., x_{22}^2,
$13+\theta$ represents the value of the 4th basic variable i.e., s_1^2,
$9+0\theta$ represents the value of the 5th basic variable i.e., s_2^2,
$1-\theta$ represents the value of the 6th variable i.e., s_4^2 and
the value of the 7th variable i.e., $s_{22}^2 \rightarrow \infty$ in a more-for-less solution of crisp linear fractional transportation problem with mixed constraints, represented by Table 5.4.
For example, substituting $\theta = 1$, the obtained values of basic variables in a more-for-less solution are

$$x_{13}^2 = 6,$$

$$x_{21}^2 = 9,$$

$$x_{22}^2 = 15,$$

$$s_1^2 = 14,$$

$$s_2^2 = 9,$$

$$s_4^2 = 0,$$

$s_{22} \rightarrow \infty$ and the corresponding cost is $\frac{5x_{11}^2+4x_{12}^2+2x_{13}^2+6x_{21}^2+5x_{22}^2+3x_{23}^2+8x_{31}^2+9x_{32}^2+4x_{33}^2}{6x_{11}^2+3x_{12}^2+4x_{13}^2+7x_{21}^2+4x_{22}^2+2x_{23}^2+6x_{31}^2+5x_{32}^2+2x_{33}^2} = 0.9592.$

(ii) For any value θ such that $0 < \theta \leq 1$,
$5+\theta$ represents the value of the 1st basic variable i.e.,
x_{13}^2, $8+0\theta$ represents the value of the 2nd basic variable i.e., x_{21}^2,
$15+\theta$ represents the value of the 3rd variable i.e., x_{22}^2,
$13+\theta$ represents the value of the 4th basic variable i.e., s_1^2,
$9+0\theta$ represents the value of the 5th basic variable i.e., s_2^2,
$1-\theta$ represents the value of the 6th variable i.e., s_4^2 and
the value of the 7th variable i.e., $s_{22}^2 \rightarrow \infty$ in a more-for-less solution of crisp linear fractional transportation problem with mixed constraints, represented by Table 5.4.

For example, substituting $\theta = 1$, the obtained values of basic variables in a more-for-less solution are

$$x_{13}^2 = 6,$$

$$x_{21}^2 = 8,$$

$$x_{22}^2 = 16,$$

$$s_1^2 = 14,$$

$$s_2^2 = 9,$$

$$s_4^2 = 0,$$

$s_{22} \rightarrow \infty$ and the corresponding cost is $\frac{5x_{11}^2+4x_{12}^2+2x_{13}^2+6x_{21}^2+5x_{22}^2+3x_{23}^2+8x_{31}^2+9x_{32}^2+4x_{33}^2}{6x_{11}^2+3x_{12}^2+4x_{13}^2+7x_{21}^2+4x_{22}^2+2x_{23}^2+6x_{31}^2+5x_{32}^2+2x_{33}^2} = 0.9722.$

(iii) For any value θ such that $\theta > 0$,
$5 + \theta$ represents the value of the 1st basic variable i.e., x_{13}^2,
$8 + 0\theta$ represents the value of the 2nd basic variable i.e., x_{21}^2,
$15 + 0\theta$ represents the value of the 3rd variable i.e., x_{22}^2,
$13 + 0\theta$ represents the value of the 4th basic variable i.e., s_1^2,
$9 + 0\theta$ represents the value of the 5th basic variable i.e., s_2^2,
$1 + 0\theta$ represents the value of the 6th variable i.e., s_4^2 and
the value of the 7th variable i.e., $s_{22}^2 \rightarrow \infty$ in a more-for-less solution of crisp linear fractional transportation problem with mixed constraints, represented by Table 5.4.

For example, considering θ tends to ∞, the obtained values of basic variables in a more-for-less solution are

$$x_{13}^2 \rightarrow \infty,$$

$$x_{21}^2 = 8,$$

$$x_{22}^2 = 15,$$

$$s_1^2 = 13,$$

$$s_2^2 = 9,$$

$$s_4^2 = 1,$$

$s_{22} \to \infty$ and the corresponding cost is $\lim\limits_{x_{13}\to\infty} \frac{5x_{11}^2+4x_{12}^2+2x_{13}^2+6x_{21}^2+5x_{22}^2+3x_{23}^2+8x_{31}^2+9x_{32}^2+4x_{33}^2}{6x_{11}^2+3x_{12}^2+4x_{13}^2+7x_{21}^2+4x_{22}^2+2x_{23}^2+6x_{31}^2+5x_{32}^2+2x_{33}^2} = \frac{2}{4} = 0.5.$

(iv) For any value θ such that $\theta > 0$,

$5 + 0\theta$ represents the value of the 1st basic variable i.e., x_{13}^2,
$8 + \theta$ represents the value of the 2nd basic variable i.e., x_{21}^2,
$15 + 0\theta$ represents the value of the 3rd variable i.e., x_{22}^2,
$13 + 0\theta$ represents the value of the 4th basic variable i.e., s_1^2,
$9 + 0\theta$ represents the value of the 5th basic variable i.e., s_2^2,
$1 + 0\theta$ represents the value of the 6th variable i.e., s_4^2 and
the value of the 7th variable i.e., $s_{22}^2 \to \infty$ in a more-for-less solution of crisp linear fractional transportation problem with mixed constraints, represented by Table 5.4.

For example, considering θ tends to ∞, the obtained values of basic variables in a more-for-less solution are

$$x_{13}^2 = 5,$$

$$x_{21}^2 \to \infty,$$

$$x_{22}^2 = 15,$$

$$s_1^2 = 13,$$

$$s_2^2 = 9,$$

$$s_4^2 = 1,$$

$s_{22} \to \infty$ and the corresponding cost is $\lim\limits_{x_{21}\to\infty} \frac{5x_{11}^2+4x_{12}^2+2x_{13}^2+6x_{21}^2+5x_{22}^2+3x_{23}^2+8x_{31}^2+9x_{32}^2+4x_{33}^2}{6x_{11}^2+3x_{12}^2+4x_{13}^2+7x_{21}^2+4x_{22}^2+2x_{23}^2+6x_{31}^2+5x_{32}^2+2x_{33}^2} = \frac{6}{7} = 0.8571.$

(v) For any value θ such that $\theta > 0$,

$5 + 0\theta$ represents the value of the 1st basic variable i.e., x_{13}^2, $8 + \theta$ represents the value of the 2nd basic variable i.e., x_{21}^2,
$15 + 0\theta$ represents the value of the 3rd variable i.e., x_{22}^2,
$13 + \theta$ represents the value of the 4th basic variable i.e., s_1^2,
$9 + \theta$ represents the value of the 5th basic variable i.e., s_2^2,
$1 + 0\theta$ represents the value of the 6th variable i.e., s_4^2 and

the value of the 7th variable i.e., $s_{22}^2 \to \infty$ in a more-for-less solution of crisp linear fractional transportation problem with mixed constraints, represented by Table 5.4.

For example, considering θ tends to ∞, the obtained values of basic variables in a more-for-less solution are

$$x_{13}^2 = 5,$$

$$x_{21}^2 \to \infty,$$

$$x_{22}^2 = 15,$$

$$s_1^2 \to \infty,$$

$$s_2^2 \to \infty,$$

$$s_4^2 = 1,$$

$s_{22} \to \infty$ and the corresponding cost is $\lim_{x_{21} \to \infty} \frac{5x_{11}^2+4x_{12}^2+2x_{13}^2+6x_{21}^2+5x_{22}^2+3x_{23}^2+8x_{31}^2+9x_{32}^2+4x_{33}^2}{6x_{11}^2+3x_{12}^2+4x_{13}^2+7x_{21}^2+4x_{22}^2+2x_{23}^2+6x_{31}^2+5x_{32}^2+2x_{33}^2} = \frac{6}{7} = 0.8571.$

5.8.2 *All More-For-Less Solutions of the Second Existing Problem*

Gupta et al. [4] considered the crisp linear fractional transportation problem with mixed constraints, represented by Table 5.5, to illustrate their proposed method. In this section, all more-for-less solutions (if exists) of the same problem are obtained by the proposed Mehar method-IV.

Using the proposed Mehar method-IV, all more-for-less solutions (if exists) of the existing crisp linear fractional transportation problem with mixed constraints, represented by Table 5.5, can be obtained as follows.

Step 1: According to **Step 1** of the proposed Mehar method-IV, there is a need to use **Step 1.1** to **Step 1.5** to check that for the crisp transportation problem with mixed constraints, represented by Table 5.5, at least one more-for-less solution will exist or not.

Step 1.1: According to **Step 1.1** of the proposed Mehar method-IV, there is a need to transform the symmetric triangular intuitionistic fuzzy linear fractional transportation problem with mixed constraints into its equivalent crisp linear

Table 5.5 Second crisp linear fractional transportation problem with mixed constraints [4]

	D_1	D_2	D_3	Availability
S_1	$c_{11}^2 = 5$ $d_{11}^2 = 6$	$c_{12}^2 = 4$ $d_{12}^2 = 3$	$c_{13}^2 = 2$ $d_{13}^2 = 4$	≥ 5
S_2	$c_{21}^2 = 6$ $d_{21}^2 = 7$	$c_{22}^2 = 5$ $d_{22}^2 = 4$	$c_{23}^2 = 3$ $d_{23}^2 = 2$	≥ 10
S_3	$c_{31}^2 = 8$ $d_{31}^2 = 6$	$c_{32}^2 = 9$ $d_{32}^2 = 5$	$c_{33}^2 = 4$ $d_{33}^2 = 2$	≤ 9
Demand	≥ 8	≥ 15	≤ 6	

fractional transportation problem with mixed constraints. Since, the linear fractional transportation problem with mixed constraints, represented by Table 5.5, is a crisp linear fractional transportation problem with mixed constraints. So, there is no need to apply this step.

Step 1.2: According to **Step 1.2** of the proposed Mehar method-IV, there is a need to find an optimal solution of the crisp linear programming problem ($P5.16$) which is obtained by substituting $P_{11} = P_{12} = P_{21} = P_{22} = M$, $Q_{11} = Q_{12} = Q_{21} = Q_{22} = 1$, $P_{13} = P_{23} = P_{31} = P_{32} = P_{33} = Q_{13} = Q_{23} = Q_{31} = Q_{32} = Q_{33} = 0$ as on solving the problems ($P5.17^{112}$), ($P5.17^{122}$), ($P5.17^{212}$) and ($P5.17^{222}$), an unbounded optimal solution is obtained. While, on solving the problems ($P5.17^{132}$), ($P5.17^{232}$), ($P5.17^{312}$), ($P5.17^{322}$) and ($P5.17^{332}$), a bounded optimal solution is obtained.

Problem ($\boldsymbol{P}5.16$)

Maximize (k)

Subject to

$$- 5u_1^2 - 10u_2^2 - 9u_3^2 - 8v_1^2 - 15v_2^2 - 6v_3^2 - Mw_{11}^2$$
$$- Mw_{12}^2 - Mw_{21}^2 - Mw_{22}^2 \leq 0,$$

$$u_1^2 + v_1^2 + w_{11}^2 + 6k \leq 5,$$

$$u_1^2 + v_2^2 + w_{12}^2 + 3k \leq 4,$$

$$u_1^2 + v_3^2 + 4k \leq 2,$$

$$u_2^2 + v_1^2 + w_{21}^2 + 7k \leq 6,$$

$$u_2^2 + v_2^2 + w_{22}^2 + 4k \leq 5,$$

$$u_2^2 + v_3^2 + 2k \leq 3,$$

$$u_3^2 + v_1^2 + 6k \leq 8,$$

$$u_3^2 + v_2^2 + 5k \leq 9,$$

$$u_3^2 + v_3^2 + 2k \leq 4,$$

$$u_1^2, u_2^2, v_1^2, v_2^2 \geq 0,$$

$$u_3^2, v_3^2, w_{11}^2, w_{12}^2, w_{21}^2, w_{22}^2 \leq 0,$$

k is unrestricted in sign.

Problem $\left(P5.17^{112}\right)$
Maximize $\left(x_{11}^2\right)$
Subject to

$$x_{11}^2 + x_{12}^2 + x_{13}^2 \geq 5,$$

$$x_{21}^2 + x_{22}^2 + x_{23}^2 \geq 10,$$

$$x_{31}^2 + x_{32}^2 + x_{33}^2 \leq 9,$$

$$x_{11}^2 + x_{21}^2 + x_{31}^2 \geq 8,$$

$$x_{12}^2 + x_{22}^2 + x_{32}^2 \geq 15,$$

$$x_{13}^2 + x_{23}^2 + x_{33}^2 \leq 6,$$

$$x_{ij} \geq 0, i = 1, 2, 3;\ j = 1, 2, 3.$$

Problem $\left(P5.17^{122}\right)$
Maximize $\left(x_{12}^2\right)$
Subject to
Constraints of the problem $(P5.17^{112})$.

Problem $\left(P5.17^{132}\right)$
Maximize $\left(x_{13}^2\right)$
Subject to

Constraints of the problem ($P5.17^{112}$).

Problem $\left(P5.17^{212}\right)$

Maximize $\left(x_{21}^2\right)$

Subject to

Constraints of the problem ($P5.17^{112}$).

Problem $\left(P5.17^{222}\right)$

Maximize $\left(x_{22}^2\right)$

Subject to

Constraints of the problem ($P5.17^{112}$).

Problem $\left(P5.17^{232}\right)$

Maximize $\left(x_{23}^2\right)$

Subject to

Constraints of the problem ($P5.17^{112}$).

Problem $\left(P5.17^{312}\right)$

Maximize $\left(x_{31}^2\right)$

Subject to

Constraints of the problem ($P5.17^{112}$).

Problem $\left(P5.17^{322}\right)$

Maximize $\left(x_{32}^2\right)$

Subject to

Constraints of the problem ($P5.17^{112}$).

Problem $\left(P5.17^{332}\right)$

Maximize $\left(x_{33}^2\right)$

Subject to

Constraints of the problem ($P5.17^{112}$).

It can be easily verified that on solving the problem ($P5.16$), the obtained optimal solution is $u_1^2 = 0, u_2^2 = \frac{1}{6}, u_3^2 = 0, v_1^2 = 0, v_2^2 = \frac{3}{2}, v_3^2 = 0, w_{11}^2 = 0, w_{12}^2 = 0, w_{21}^2 = 0, w_{22}^2 = 0$

Step 1.3: It is obvious that $5u_1^2 + 10u_2^2 + 9u_3^2 + 8v_1^2 + 15v_2^2 + 6v_3^2 + Mw_{11}^2 + Mw_{12}^2 + Mw_{21}^2 + Mw_{22}^2 \neq 0$. So, according to Case (ii) of **Step 1.3** of the proposed Mehar method-IV, there does not exist a more-for-less solution of the crisp transportation problem with mixed constraints, represented by Table 5.5, as it is not possible to increase the total optimal supplied quantity of the product.

5.8.3 All More-For-Less Solutions of the Considered Problem

In this section, all more-for-less solutions (if exists) of the symmetric triangular intuitionistic fuzzy linear fractional transportation problem with mixed constraints, represented by Table 5.6, are obtained by the proposed Mehar method-IV.

Using the proposed Mehar method-IV, all more-for-less solutions (if exists) of the symmetric triangular intuitionistic fuzzy linear fractional transportation problem with mixed constraints, represented by Table 5.6, can be obtained as follows.

Table 5.6 Symmetric triangular intuitionistic fuzzy linear fractional transportation problem with mixed constraints

	D_1	D_2	D_3	Availability
S_1	$\tilde{c}_{11}$ $\tilde{d}_{11}$	$\tilde{c}_{12}$ $\tilde{d}_{12}$	$\tilde{c}_{13}$ $\tilde{d}_{13}$	$=_{Mag}\ \tilde{a}_1$
S_2	$\tilde{c}_{21}$ $\tilde{d}_{21}$	$\tilde{c}_{22}$ $\tilde{d}_{22}$	$\tilde{c}_{23}$ $\tilde{d}_{23}$	$\geq_{Mag}\ \tilde{a}_2$
S_3	$\tilde{c}_{31}$ $\tilde{d}_{31}$	$\tilde{c}_{32}$ $\tilde{d}_{32}$	$\tilde{c}_{33}$ $\tilde{d}_{33}$	$\leq_{Mag}\ \tilde{a}_3$
Demand	$=_{Mag}\ \tilde{b}_1$	$\geq_{Mag}\ \tilde{b}_2$	$\leq_{Mag}\ \tilde{b}_3$	

where,

(i) $\tilde{c}_{11} = (5, 3(1-\alpha), 3(1-\alpha); 5, 4(1-\alpha), 4(1-\alpha))$, $\tilde{c}_{12} = (4, 2(1-\alpha), 2(1-\alpha); 4, 3(1-\alpha), 3(1-\alpha))$, $\tilde{c}_{13} = (2, 1(1-\alpha), 1(1-\alpha); 2, 2(1-\alpha), 2(1-\alpha))$, $\tilde{c}_{21} = (6, 1(1-\alpha), 1(1-\alpha); 6, 5(1-\alpha), 5(1-\alpha))$, $\tilde{c}_{22} = (5, 1(1-\alpha), 1(1-\alpha); 5, 3(1-\alpha), 3(1-\alpha))$, $\tilde{c}_{23} = (3, 1(1-\alpha), 1(1-\alpha); 3, 2(1-\alpha), 2(1-\alpha))$, $\tilde{c}_{31} = (8, 1(1-\alpha), 1(1-\alpha); 8, 3(1-\alpha), 3(1-\alpha))$, $\tilde{c}_{32} = (9, 3(1-\alpha), 3(1-\alpha); 9, 5(1-\alpha), 5(1-\alpha))$, $\tilde{c}_{33} = (4, 1(1-\alpha), 1(1-\alpha); 4, 3(1-\alpha), 3(1-\alpha))$

(ii) $\tilde{d}_{11} = (6, 2(1-\alpha), 2(1-\alpha); 6, 4(1-\alpha), 4(1-\alpha))$, $\tilde{d}_{12} = (3, 1, 1(1-\alpha); 3, 2(1-\alpha), 2(1-\alpha))$, $\tilde{d}_{13} = (4, 1(1-\alpha), 1(1-\alpha); 4, 2(1-\alpha), 2(1-\alpha))$, $\tilde{d}_{21} = (7, 2(1-\alpha), 2(1-\alpha); 7, 3(1-\alpha), 3(1-\alpha))$, $\tilde{d}_{22} = (4, 1(1-\alpha), 1(1-\alpha); 4, 2(1-\alpha), 2(1-\alpha))$, $\tilde{d}_{23} = (2, 1(1-\alpha), 1(1-\alpha); 2, 1(1-\alpha), 1(1-\alpha))$, $\tilde{d}_{31} = (6, 2(1-\alpha), 2(1-\alpha); 6, 3(1-\alpha), 3(1-\alpha))$, $\tilde{d}_{32} = (5, 3(1-\alpha), 3(1-\alpha); 5, 4(1-\alpha), 4(1-\alpha))$, $\tilde{d}_{33} = (2, 1(1-\alpha), 1(1-\alpha); 2, 2(1-\alpha), 2(1-\alpha))$

(iii) $\tilde{a}_1 = (5, 1(1-\alpha), 1(1-\alpha); 5, 4(1-\alpha), 4(1-\alpha))$, $\tilde{a}_2 = (10, 3(1-\alpha), 3(1-\alpha); 10, 9(1-\alpha), 9(1-\alpha))$, $\tilde{a}_3 = (9, 3(1-\alpha), 3(1-\alpha); 9, 6(1-\alpha), 6(1-\alpha))$

(iv) $\tilde{b}_1 = (8, 4(1-\alpha), 4(1-\alpha); 8, 7(1-\alpha), 7(1-\alpha))$, $\tilde{b}_2 = (15, 10(1-\alpha), 10(1-\alpha); 15, 13(1-\alpha), 13(1-\alpha))$, $\tilde{b}_3 = (6, 2(1-\alpha), 2(1-\alpha); 6, 4(1-\alpha), 4(1-\alpha))$

Step 1: According to **Step 1** of the proposed Mehar method-IV, there is a need to use **Step 1.1** to **Step 1.5** to check that for the symmetric triangular intuitionistic fuzzy linear fractional transportation problem with mixed constraints, represented by Table 5.6, at least one more-for-less solution will exist or not.

Step 1.1: According to **Step 1.1** of the proposed Mehar method-IV, the symmetric triangular intuitionistic fuzzy linear fractional transportation problem with mixed constraints, represented by Table 5.6, can be transformed into its equivalent crisp linear fractional transportation problem with mixed constraints represented by Table 5.4.

It is obvious from Section 5.8.1 that there exists at least one more-for-less solution of the crisp linear fractional transportation problem with mixed constraints, represented by Table 5.4. Hence, for the symmetric triangular intuitionistic fuzzy linear fractional transportation problem with mixed constraints, represented by Table 5.6, at least one more-for-less solution will also exist.

Step 2: It is obvious from Section 5.8.1 that on using **Step 2.1** to **Step 2.9**, the following more-for-less solutions are obtained.

(i) For any value θ such that $0 < \theta \leq 1$,
$5 + \theta$ represents the value of the 1st basic variable i.e., x_{13}^2,
$8 + \theta$ represents the value of the 2nd basic variable i.e., x_{21}^2,
$15 + 0\theta$ represents the value of the 3rd variable i.e., x_{22}^2,
$13 + \theta$ represents the value of the 4th basic variable i.e., s_1^2,
$9 + 0\theta$ represents the value of the 5th basic variable i.e., s_2^2,
$1 - \theta$ represents the value of the 6th variable i.e., s_4^2 and
the value of the 7th variable i.e., $s_{22}^2 \to \infty$ in a more-for-less solution of crisp linear fractional transportation problem with mixed constraints, represented by Table 5.4.

For example, substituting $\theta = 1$, the obtained values of basic variables in a more-for-less solution are

$$x_{13}^2 = 6,$$

$$x_{21}^2 = 9,$$

$$x_{22}^2 = 15,$$

$$s_1^2 = 14,$$

$$s_2^2 = 9,$$

$$s_4^2 = 0,$$

$s_{22} \to \infty$ and the corresponding cost is $\frac{5x_{11}^2+4x_{12}^2+2x_{13}^2+6x_{21}^2+5x_{22}^2+3x_{23}^2+8x_{31}^2+9x_{32}^2+4x_{33}^2}{6x_{11}^2+3x_{12}^2+4x_{13}^2+7x_{21}^2+4x_{22}^2+2x_{23}^2+6x_{31}^2+5x_{32}^2+2x_{33}^2} = 0.9592$

(ii) For any value θ such that $0 < \theta \leq 1$,
$5 + \theta$ represents the value of the 1st basic variable i.e., x_{13}^2,
$8 + 0\theta$ represents the value of the 2nd basic variable i.e., x_{21}^2,
$15 + \theta$ represents the value of the 3rd variable i.e., x_{22}^2,
$13 + \theta$ represents the value of the 4th basic variable i.e., s_1^2,
$9 + 0\theta$ represents the value of the 5th basic variable i.e., s_2^2,
$1 - \theta$ represents the value of the 6th variable i.e., s_4^2 and
the value of the 7th variable i.e., $s_{22}^2 \to \infty$ in a more-for-less solution of crisp linear fractional transportation problem with mixed constraints, represented by Table 5.4.

For example, substituting $\theta = 1$, the obtained values of basic variables in a more-for-less solution are

$$x_{13}^2 = 6,$$

$$x_{21}^2 = 8,$$

$$x_{22}^2 = 16,$$

$$s_1^2 = 14,$$

$$s_2^2 = 9,$$

$$s_4^2 = 0,$$

$s_{22} \rightarrow \infty$ and the corresponding cost is $\frac{5x_{11}^2+4x_{12}^2+2x_{13}^2+6x_{21}^2+5x_{22}^2+3x_{23}^2+8x_{31}^2+9x_{32}^2+4x_{33}^2}{6x_{11}^2+3x_{12}^2+4x_{13}^2+7x_{21}^2+4x_{22}^2+2x_{23}^2+6x_{31}^2+5x_{32}^2+2x_{33}^2} = 0.9722$

(iii) For any value θ such that $\theta > 0$,
$5+\theta$ represents the value of the 1st basic variable i.e., x_{13}^2,
$8+0\theta$ represents the value of the 2nd basic variable i.e., x_{21}^2,
$15+0\theta$ represents the value of the 3rd variable i.e., x_{22}^2,
$13+0\theta$ represents the value of the 4th basic variable i.e., s_1^2,
$9+0\theta$ represents the value of the 5th basic variable i.e., s_2^2,
$1+0\theta$ represents the value of the 6th variable i.e., s_4^2 and
the value of the 7th variable i.e., $s_{22}^2 \rightarrow \infty$ in a more-for-less solution of crisp linear fractional transportation problem with mixed constraints, represented by Table 5.4.

For example, considering θ tends to ∞, the obtained values of basic variables in a more-for-less solution are

$$x_{13}^2 \rightarrow \infty,$$

$$x_{21}^2 = 8,$$

$$x_{22}^2 = 15,$$

$$s_1^2 = 13,$$

$$s_2^2 = 9,$$

$$s_4^2 = 1,$$

$s_{22} \rightarrow \infty$ and the corresponding cost is $\lim\limits_{x_{13}\rightarrow\infty} \frac{5x_{11}^2+4x_{12}^2+2x_{13}^2+6x_{21}^2+5x_{22}^2+3x_{23}^2+8x_{31}^2+9x_{32}^2+4x_{33}^2}{6x_{11}^2+3x_{12}^2+4x_{13}^2+7x_{21}^2+4x_{22}^2+2x_{23}^2+6x_{31}^2+5x_{32}^2+2x_{33}^2} = \frac{2}{4} = 0.5$

(iv) For any value θ such that $\theta > 0$,

$5 + 0\theta$ represents the value of the 1st basic variable i.e., x_{13}^2,
$8 + \theta$ represents the value of the 2nd basic variable i.e., x_{21}^2,
$15 + 0\theta$ represents the value of the 3rd variable i.e., x_{22}^2,
$13 + 0\theta$ represents the value of the 4th basic variable i.e., s_1^2,
$9 + 0\theta$ represents the value of the 5th basic variable i.e., s_2^2,
$1 + 0\theta$ represents the value of the 6th variable i.e., s_4^2 and
the value of the 7th variable i.e., $s_{22}^2 \to \infty$ in a more-for-less solution of crisp linear fractional transportation problem with mixed constraints, represented by Table 5.4.

For example, considering θ tends to ∞, the obtained values of basic variables in a more-for-less solution are

$$x_{13}^2 = 5,$$

$$x_{21}^2 \to \infty,$$

$$x_{22}^2 = 15,$$

$$s_1^2 = 13,$$

$$s_2^2 = 9,$$

$$s_4^2 = 1,$$

$s_{22} \to \infty$ and the corresponding cost is $\lim\limits_{x_{21}\to\infty} \frac{5x_{11}^2+4x_{12}^2+2x_{13}^2+6x_{21}^2+5x_{22}^2+3x_{23}^2+8x_{31}^2+9x_{32}^2+4x_{33}^2}{6x_{11}^2+3x_{12}^2+4x_{13}^2+7x_{21}^2+4x_{22}^2+2x_{23}^2+6x_{31}^2+5x_{32}^2+2x_{33}^2} = \frac{6}{7} = 0.8571.$

(v) For any value θ such that $\theta > 0$,

$5 + 0\theta$ represents the value of the 1st basic variable i.e., x_{13}^2,
$8 + \theta$ represents the value of the 2nd basic variable i.e., ,
$15 + 0\theta$ represents the value of the 3rd variable i.e., x_{22}^2,
$13 + \theta$ represents the value of the 4th basic variable i.e., s_1^2,
$9 + \theta$ represents the value of the 5th basic variable i.e., s_2^2,
$1 + 0\theta$ represents the value of the 6th variable i.e., s_4^2 and
the value of the 7th variable i.e., $s_{22}^2 \to \infty$ in a more-for-less solution of crisp linear fractional transportation problem with mixed constraints, represented by Table 5.4.

For example, considering θ tends to ∞, the obtained values of basic variables in a more-for-less solution are

$$x_{13}^2 = 5,$$

$$x_{21}^2 \to \infty,$$

$$x_{22}^2 = 15,$$

$$s_1^2 \to \infty,$$

$$s_2^2 \to \infty,$$

$$s_4^2 = 1,$$

$s_{22} \to \infty$ and the corresponding cost is $\lim_{x_{21}\to\infty} \frac{5x_{11}^2+4x_{12}^2+2x_{13}^2+6x_{21}^2+5x_{22}^2+3x_{23}^2+8x_{31}^2+9x_{32}^2+4x_{33}^2}{6x_{11}^2+3x_{12}^2+4x_{13}^2+7x_{21}^2+4x_{22}^2+2x_{23}^2+6x_{31}^2+5x_{32}^2+2x_{33}^2} = \frac{6}{7} = 0.8571$

Step 2.10: According to **Step 2.10** of the proposed Mehar method-IV,

(i) For any value θ such that $0 < \theta \leq 1$,
$(5+\theta, \delta_{135}(1-\alpha), \delta_{135}(1-\alpha); 5+\theta, \delta_{136}(1-\alpha), \delta_{136}(1-\alpha))$ represents the value of the 1st basic variable i.e., $\tilde{x}_{13}$,
$(8+\theta, \delta_{215}(1-\alpha), \delta_{215}(1-\alpha); 8+\theta, \delta_{216}(1-\alpha), \delta_{216}(1-\alpha))$ represents the value of the 2nd basic variable i.e., $\tilde{x}_{21}$,
$(15+0\theta, \delta_{225}(1-\alpha), \delta_{225}(1-\alpha); 15+0\theta, \delta_{226}(1-\alpha), \delta_{226}(1-\alpha))$ represents the value of the 3rd variable i.e., $\tilde{x}_{22}$,
$(13+\theta, \gamma_{11}(1-\alpha), \gamma_{11}(1-\alpha); 13+\theta, \gamma_{12}(1-\alpha), \gamma_{12}(1-\alpha))$ represents the value of the 4th basic variable i.e., $\tilde{s}_1$,
$(9+0\theta, \gamma_{21}(1-\alpha), \gamma_{21}(1-\alpha); 9+0\theta, \gamma_{22}(1-\alpha), \gamma_{22}(1-\alpha))$ represents the value of the 5th basic variable i.e., $\tilde{s}_2$,
$(1-\theta, \gamma_{41}(1-\alpha), \gamma_{41}(1-\alpha); 1-\theta, \gamma_{42}(1-\alpha), \gamma_{42}(1-\alpha))$ represents the value of the 6th variable i.e., $\tilde{s}_4$ and
the value of the 7th variable i.e., $s_{22}^2 \to \infty$. in a more-for-less solution of the symmetric triangular fuzzy transportation problem with mixed constraints, represented by Table 5.6.
For example,
(a) Substituting $\theta = 1$, $\delta_{ij5} = \delta_{ij6} = 1 \forall i, j$ and $\gamma_{k1} = \gamma_{k2} = 1; k = 1, 2, 4$, the obtained values of basic variables in a more-for-less solution are $\tilde{x}_{13} =_{Mag} (6, 1(1-\alpha), 1(1-\alpha); 6, 1(1-\alpha), 1(1-\alpha))$,

$$\tilde{x}_{21} =_{Mag} (9, 1(1-\alpha), 1(1-\alpha); 9, 1(1-\alpha), 1(1-\alpha)),$$

$$\tilde{x}_{22} =_{Mag} (15, 1(1-\alpha), 1(1-\alpha); 15, 1(1-\alpha), 1(1-\alpha)),$$

$$\tilde{s}_1 =_{Mag} (14, 1(1-\alpha), 1(1-\alpha); 14, 1(1-\alpha), 1(1-\alpha)),$$

$$\tilde{s}_2 =_{Mag} (9, 1(1-\alpha), 1(1-\alpha); 9, 1(1-\alpha), 1(1-\alpha)),$$

$$\tilde{s}_4 =_{Mag} (0, 1(1-\alpha), 1(1-\alpha); 0, 1(1-\alpha), 1(1-\alpha)),$$

$s_{22}^2 \to \infty$ and the corresponding symmetric triangular intuitionistic fuzzy transportation cost is $\tilde{Z} =_{Mag} \frac{(141,1(1-\alpha),1(1-\alpha);141,5(1-\alpha),5(1-\alpha))}{(147,2(1-\alpha),2(1-\alpha);147,3(1-\alpha),3(1-\alpha))}$.

(b) Substituting $\theta = 1$, $\delta_{ij5} = 2$, $\delta_{ij6} = 3 \forall i, j$ and $\gamma_{k1} = \gamma_{k2} = 1; k = 1, 2, 4$, the obtained values of basic variables in a more-for-less solution are $\tilde{x}_{13} =_{Mag} (6, 2(1-\alpha), 2(1-\alpha); 6, 3(1-\alpha), 3(1-\alpha))$,

$$\tilde{x}_{21} =_{Mag} (9, 2(1-\alpha), 2(1-\alpha); 9, 3(1-\alpha), 3(1-\alpha)),$$

$$\tilde{x}_{22} =_{Mag} (15, 2(1-\alpha), 2(1-\alpha); 15, 3(1-\alpha), 3(1-\alpha)),$$

$$\tilde{s}_1 =_{Mag} (14, 1(1-\alpha), 1(1-\alpha); 14, 1(1-\alpha), 1(1-\alpha)),$$

$$\tilde{s}_2 =_{Mag} (9, 1(1-\alpha), 1(1-\alpha); 9, 1(1-\alpha), 1(1-\alpha)),$$

$$\tilde{s}_4 =_{Mag} (0, 1(1-\alpha), 1(1-\alpha); 0, 1(1-\alpha), 1(1-\alpha)),$$

$s_{22}^2 \to \infty$ and the corresponding symmetric triangular intuitionistic fuzzy transportation cost is $\tilde{Z} =_{Mag} \frac{(141,2(1-\alpha),2(1-\alpha);141,5(1-\alpha),5(1-\alpha))}{(147,2(1-\alpha),2(1-\alpha);147,3(1-\alpha),3(1-\alpha))}$.

(ii) For any value θ such that $0 < \theta \leq 1$,

$(5+\theta, \delta_{135}(1-\alpha), \delta_{135}(1-\alpha); 5+\theta, \delta_{136}(1-\alpha), \delta_{136}(1-\alpha))$ represents the value of the 1st basic variable i.e., $\tilde{x}_{13}$,

$(8+0\theta, \delta_{215}(1-\alpha), \delta_{215}(1-\alpha); 8+0\theta, \delta_{216}(1-\alpha), \delta_{216}(1-\alpha))$ represents the value of the 2nd basic variable i.e., $\tilde{x}_{21}$,

$(15+\theta, \delta_{225}(1-\alpha), \delta_{225}(1-\alpha); 15+\theta, \delta_{226}(1-\alpha), \delta_{226}(1-\alpha))$ represents the value of the 3rd variable i.e., $\tilde{x}_{22}$,

$(13+\theta, \gamma_{11}(1-\alpha), \gamma_{11}(1-\alpha); 13+\theta, \gamma_{12}(1-\alpha), \gamma_{12}(1-\alpha))$ represents the value of the 4th basic variable i.e., $\tilde{s}_1$,

$(9+0\theta, \gamma_{21}(1-\alpha), \gamma_{21}(1-\alpha); 9+0\theta, \gamma_{22}(1-\alpha), \gamma_{22}(1-\alpha))$ represents the value of the 5th basic variable i.e., $\tilde{s}_2$,

$(1-\theta, \gamma_{41}(1-\alpha), \gamma_{41}(1-\alpha); 1-\theta, \gamma_{42}(1-\alpha), \gamma_{42}(1-\alpha))$ represents the value of the 6th variable i.e., $\tilde{s}_4$ and

the value of the 7th variable i.e., $s_{22}^2 \to \infty$ in a more-for-less solution of the symmetric triangular fuzzy transportation problem with mixed constraints, represented by Table 5.6.

For example,

(a) Substituting $\theta = 1$, $\delta_{ij5} = \delta_{ij6} = 2 \forall i, j$ and $\gamma_{k1} = \gamma_{k2} = 1; k = 1, 2, 4$, the obtained values of basic variables in a more-for-less solution are $\tilde{x}_{13} =_{Mag} (6, 2(1-\alpha), 2(1-\alpha); 6, 2(1-\alpha), 2(1-\alpha))$,

$$\tilde{x}_{21} =_{Mag} (8, 2(1-\alpha), 2(1-\alpha); 8, 2(1-\alpha), 2(1-\alpha)),$$

$$\tilde{x}_{22} =_{Mag} (16, 2(1-\alpha), 2(1-\alpha); 16, 2(1-\alpha), 2(1-\alpha)),$$

$$\tilde{s}_1 =_{Mag} (14, 1(1-\alpha), 1(1-\alpha); 14, 1(1-\alpha), 1(1-\alpha)),$$

$$\tilde{s}_2 =_{Mag} (9, 1(1-\alpha), 1(1-\alpha); 9, 1(1-\alpha), 1(1-\alpha)),$$

$$\tilde{s}_4 =_{Mag} (0, 1(1-\alpha), 1(1-\alpha); 0, 1(1-\alpha), 1(1-\alpha)),$$

$s_{22}^2 \to \infty$ and the corresponding symmetric triangular intuitionistic fuzzy transportation cost is $\tilde{Z} =_{Mag} \frac{(140,2(1-\alpha),2(1-\alpha);140,5(1-\alpha),5(1-\alpha))}{(144,2(1-\alpha),2(1-\alpha);144,3(1-\alpha),3(1-\alpha))}$.

(b) Substituting $\theta = 1$, $\delta_{ij5} = 5$, $\delta_{ij6} = 7 \forall i, j$ and $\gamma_{k1} = \gamma_{k2} = 4; k = 1, 2, 4$, the obtained values of basic variables in a more-for-less solution are $\tilde{x}_{13} =_{Mag} (6, 5(1-\alpha), 5(1-\alpha); 6, 7(1-\alpha), 7(1-\alpha))$,

$$\tilde{x}_{21} =_{Mag} (8, 5(1-\alpha), 5(1-\alpha); 8, 7(1-\alpha), 7(1-\alpha)),$$

$$\tilde{x}_{22} =_{Mag} (16, 5(1-\alpha), 5(1-\alpha); 16, 7(1-\alpha), 7(1-\alpha)),$$

$$\tilde{s}_1 =_{Mag} (14, 4(1-\alpha), 4(1-\alpha); 14, 4(1-\alpha), 4(1-\alpha)),$$

$$\tilde{s}_2 =_{Mag} (9, 4(1-\alpha), 4(1-\alpha); 9, 4(1-\alpha), 4(1-\alpha)),$$

$$\tilde{s}_4 =_{Mag} (0, 4(1-\alpha), 4(1-\alpha); 0, 4(1-\alpha), 4(1-\alpha)),$$

$s_{22}^2 \to \infty$ and the corresponding symmetric triangular intuitionistic fuzzy transportation cost is $\tilde{Z} =_{Mag} \frac{(140,5(1-\alpha),5(1-\alpha);140,7(1-\alpha),7(1-\alpha))}{(144,5(1-\alpha),5(1-\alpha);144,7(1-\alpha),7(1-\alpha))}$.

(iii) For any value θ such that $\theta > 0$,

$(5+\theta, \delta_{135}(1-\alpha), \delta_{135}(1-\alpha); 5+\theta, \delta_{136}(1-\alpha), \delta_{136}(1-\alpha))$ represents the value of the 1st basic variable i.e., $\tilde{x}_{13}$,

$(8+0\theta, \delta_{215}(1-\alpha), \delta_{215}(1-\alpha); 8+0\theta, \delta_{216}(1-\alpha), \delta_{216}(1-\alpha))$ represents the value of the 2nd basic variable i.e., $\tilde{x}_{21}$,

$(15+0\theta, \delta_{225}(1-\alpha), \delta_{225}(1-\alpha); 15+0\theta, \delta_{226}(1-\alpha), \delta_{226}(1-\alpha))$ represents the value of the 3rd variable i.e., $\tilde{x}_{22}$,

$(13+0\theta, \gamma_{11}(1-\alpha), \gamma_{11}(1-\alpha); 13+0\theta, \gamma_{12}(1-\alpha), \gamma_{12}(1-\alpha))$ represents the value of the 4th basic variable i.e., $\tilde{s}_1$, .

$(9+0\theta, \gamma_{21}(1-\alpha), \gamma_{21}(1-\alpha); 9+0\theta, \gamma_{22}(1-\alpha), \gamma_{22}(1-\alpha))$ represents the value of the 5th basic variable i.e., $\tilde{s}_2$,

$(1+0\theta, \gamma_{41}(1-\alpha), \gamma_{41}(1-\alpha); 1+0\theta, \gamma_{42}(1-\alpha), \gamma_{42}(1-\alpha))$ represents the value of the 6th variable i.e., $\tilde{s}_4$ and

the value of the 7th variable i.e., $s_{22}^2 \to \infty$ in a more-for-less solution of the symmetric triangular fuzzy transportation problem with mixed constraints, represented by Table 5.6.

For example,

(a) Considering θ tends to ∞, $\delta_{ij5} = \delta_{ij6} = 3 \forall i, j$ and $\gamma_{k1} = \gamma_{k2} = 1; k = 1, 2, 4$, the obtained values of basic variables in a more-for-less solution are $\tilde{x}_{13} \to \infty$,

$$\tilde{x}_{21} =_{Mag} (8, 3(1-\alpha), 3(1-\alpha); 8, 3(1-\alpha), 3(1-\alpha)),$$

$$\tilde{x}_{22} =_{Mag} (15, 3(1-\alpha), 3(1-\alpha); 15, 3(1-\alpha), 3(1-\alpha)),$$

$$\tilde{s}_1 =_{Mag} (13, 1(1-\alpha), 1(1-\alpha); 13, 1(1-\alpha), 1(1-\alpha)),$$

$$\tilde{s}_2 =_{Mag} (9, 1(1-\alpha), 1(1-\alpha); 9, 1(1-\alpha), 1(1-\alpha)),$$

$$\tilde{s}_4 =_{Mag} (1, 1(1-\alpha), 1(1-\alpha); 1, 1(1-\alpha), 1(1-\alpha)),$$

$s_{22}^2 \to \infty$ and the corresponding symmetric triangular intuitionistic fuzzy transportation cost is $\tilde{Z} =_{Mag} \frac{(2,3(1-\alpha),3(1-\alpha);2,5(1-\alpha),5(1-\alpha))}{(4,3(1-\alpha),3(1-\alpha);4,3(1-\alpha),3(1-\alpha))}$.

(b) Considering θ tends to ∞, $\delta_{ij5} = 1, \delta_{ij6} = 2 \forall i, j$ and $\gamma_{k1} = \gamma_{k2} = 1; k = 1, 2, 4$, the obtained values of basic variables in a more-for-less solution are $\tilde{x}_{13} \to \infty$,

$$\tilde{x}_{21} =_{Mag} (8, 1(1-\alpha), 1(1-\alpha); 8, 2(1-\alpha), 2(1-\alpha)),$$

$$\tilde{x}_{22} =_{Mag} (15, 1(1-\alpha), 1(1-\alpha); 15, 2(1-\alpha), 2(1-\alpha)),$$

$$\tilde{s}_1 =_{Mag} (13, 1(1-\alpha), 1(1-\alpha); 13, 1(1-\alpha), 1(1-\alpha)),$$

$$\tilde{s}_2 =_{Mag} (9, 1(1-\alpha), 1(1-\alpha); 9, 1(1-\alpha), 1(1-\alpha)),$$

$$\tilde{s}_4 =_{Mag} (1, 1(1-\alpha), 1(1-\alpha); 1, 1(1-\alpha), 1(1-\alpha)),$$

$s_{22}^2 \to \infty$ and the corresponding symmetric triangular intuitionistic fuzzy transportation cost is $\tilde{Z} =_{Mag} \frac{(2,1(1-\alpha),1(1-\alpha);2,5(1-\alpha),5(1-\alpha))}{(4,2(1-\alpha),2(1-\alpha);4,3(1-\alpha),3(1-\alpha))}$.

(iv) For any value θ such that $\theta > 0$,
$(5 + 0\theta, \delta_{135}(1-\alpha), \delta_{135}(1-\alpha); 5 + 0\theta, \delta_{136}(1-\alpha), \delta_{136}(1-\alpha))$ represents the value of the 1st basic variable i.e., $\tilde{x}_{13}$,
$(8 + \theta, \delta_{215}(1-\alpha), \delta_{215}(1-\alpha); 8 + \theta, \delta_{216}(1-\alpha), \delta_{216}(1-\alpha))$ represents the value of the 2nd basic variable i.e., $\tilde{x}_{21}$,
$(15 + 0\theta, \delta_{225}(1-\alpha), \delta_{225}(1-\alpha); 15 + 0\theta, \delta_{226}(1-\alpha), \delta_{226}(1-\alpha))$ represents the value of the 3rd variable i.e., $\tilde{x}_{22}$,
$(13 + 0\theta, \gamma_{11}(1-\alpha), \gamma_{11}(1-\alpha); 13 + 0\theta, \gamma_{12}(1-\alpha), \gamma_{12}(1-\alpha))$ represents the value of the 4th basic variable i.e., $\tilde{s}_1$,
$(9 + 0\theta, \gamma_{21}(1-\alpha), \gamma_{21}(1-\alpha); 9 + 0\theta, \gamma_{22}(1-\alpha), \gamma_{22}(1-\alpha))$ represents the value of the 5th basic variable i.e., $\tilde{s}_2$,
$(1 + 0\theta, \gamma_{41}(1-\alpha), \gamma_{41}(1-\alpha); 1 + 0\theta, \gamma_{42}(1-\alpha), \gamma_{42}(1-\alpha))$ represents the value of the 6th variable i.e., $\tilde{s}_4$ and
the value of the 7th variable i.e., $s_{22}^2 \to \infty$ in a more-for-less solution of the symmetric triangular fuzzy transportation problem with mixed constraints, represented by Table 5.6.

For example,

(a) Considering θ tends to ∞, $\delta_{ij5} = \delta_{ij6} = 2 \forall i, j$ and $\gamma_{k1} = \gamma_{k2} = 1; k = 1, 2, 4$, the obtained values of basic variables in a more-for-less solution are $\tilde{x}_{13} =_{Mag} (5, 2(1-\alpha), 2(1-\alpha); 5, 2(1-\alpha), 2(1-\alpha))$,

$$\tilde{x}_{21} \to \infty,$$

$$\tilde{x}_{22} =_{Mag} (15, 2(1-\alpha), 2(1-\alpha); 15, 2(1-\alpha), 2(1-\alpha)),$$

$$\tilde{s}_1 =_{Mag} (13, 1(1-\alpha), 1(1-\alpha); 13, 1(1-\alpha), 1(1-\alpha)),$$

$$\tilde{s}_2 =_{Mag} (9, 1(1-\alpha), 1(1-\alpha); 9, 1(1-\alpha), 1(1-\alpha)),$$

$$\tilde{s}_4 =_{Mag} (1, 1(1-\alpha), 1(1-\alpha); 1, 1(1-\alpha), 1(1-\alpha)),$$

$s_{22}^2 \to \infty$ and the corresponding symmetric triangular intuitionistic fuzzy transportation cost is $\tilde{Z} =_{Mag} \frac{(6,2(1-\alpha),2(1-\alpha);6,5(1-\alpha),5(1-\alpha))}{(7,2(1-\alpha),2(1-\alpha);7,3(1-\alpha),3(1-\alpha))}$.

(b) Considering θ tends to ∞, $\delta_{ij5} = 1, \delta_{ij6} = 2 \forall i, j$ and $\gamma_{k1} = \gamma_{k2} = 1; k = 1, 2, 4$, the obtained values of basic variables in a more-for-less solution are $\tilde{x}_{13} =_{Mag} (5, 1(1-\alpha), 1(1-\alpha); 5, 2(1-\alpha), 2(1-\alpha))$,

$$\tilde{x}_{21} \to \infty,$$

$$\tilde{x}_{22} =_{Mag} (15, 1(1-\alpha), 1(1-\alpha); 15, 2(1-\alpha), 2(1-\alpha)),$$

$$\tilde{s}_1 =_{Mag} (13, 1(1-\alpha), 1(1-\alpha); 13, 1(1-\alpha), 1(1-\alpha)),$$

$$\tilde{s}_2 =_{Mag} (9, 1(1-\alpha), 1(1-\alpha); 9, 1(1-\alpha), 1(1-\alpha)),$$

$$\tilde{s}_4 =_{Mag} (1, 1(1-\alpha), 1(1-\alpha); 1, 1(1-\alpha), 1(1-\alpha)),$$

$s_{22}^2 \to \infty$ and the corresponding symmetric triangular intuitionistic fuzzy transportation cost is $\tilde{Z} =_{Mag} \frac{(6,1(1-\alpha),1(1-\alpha);6,5(1-\alpha),5(1-\alpha))}{(7,2(1-\alpha),2(1-\alpha);7,3(1-\alpha),3(1-\alpha))}$.

(v) For any value θ such that $\theta > 0$,

$(5+0\theta, \delta_{135}(1-\alpha), \delta_{135}(1-\alpha); 5+0\theta, \delta_{136}(1-\alpha), \delta_{136}(1-\alpha))$ represents the value of the 1st basic variable i.e., $\tilde{x}_{13}$,

$(8+\theta, \delta_{215}(1-\alpha), \delta_{215}(1-\alpha); 8+\theta, \delta_{216}(1-\alpha), \delta_{216}(1-\alpha))$ represents the value of the 2nd basic variable i.e., $\tilde{x}_{21}$,

$(15+0\theta, \delta_{225}(1-\alpha), \delta_{225}(1-\alpha); 15+0\theta, \delta_{226}(1-\alpha), \delta_{226}(1-\alpha))$ represents the value of the 3rd variable i.e., $\tilde{x}_{22}$,

$(13+\theta, \gamma_{11}(1-\alpha), \gamma_{11}(1-\alpha); 13+\theta, \gamma_{12}(1-\alpha), \gamma_{12}(1-\alpha))$ represents the value of the 4th basic variable i.e., $\tilde{s}_1$,

$(9+\theta, \gamma_{21}(1-\alpha), \gamma_{21}(1-\alpha); 9+\theta, \gamma_{22}(1-\alpha), \gamma_{22}(1-\alpha))$ represents the value of the 5th basic variable i.e., $\tilde{s}_2$,

$(1+0\theta, \gamma_{41}(1-\alpha), \gamma_{41}(1-\alpha); 1+0\theta, \gamma_{42}(1-\alpha), \gamma_{42}(1-\alpha))$ represents the value of the 6th variable i.e., $\tilde{s}_4$ and

the value of the 7th variable i.e., $s_{22}^2 \to \infty$ in a more-for-less solution of the symmetric triangular fuzzy transportation problem with mixed constraints, represented by Table 5.6.

For example,

(a) Considering θ tends to ∞, $\delta_{ij5} = 2, \delta_{ij6} = 4 \forall i, j$ and $\gamma_{k1} = \gamma_{k2} = 1; k = 1, 2, 4$, the obtained values of basic variables in a more-for-less solution are $\tilde{x}_{13} =_{Mag} (5, 2(1-\alpha), 2(1-\alpha); 5, 4(1-\alpha), 4(1-\alpha))$,

$$\tilde{x}_{21} \to \infty,$$

$$\tilde{x}_{22} =_{Mag} (15, 2(1-\alpha), 2(1-\alpha); 15, 4(1-\alpha), 4(1-\alpha)),$$

$$\tilde{s}_1 \to \infty,$$

$$\tilde{s}_2 \to \infty,$$

$$\tilde{s}_4 =_{Mag} (1, 1(1-\alpha), 1(1-\alpha); 1, 1(1-\alpha), 1(1-\alpha)),$$

$s_{22}^2 \to \infty$ and the corresponding symmetric triangular intuitionistic fuzzy transportation cost is $\tilde{Z} =_{Mag} \frac{(6,2(1-\alpha),2(1-\alpha);6,5(1-\alpha),5(1-\alpha))}{(7,2(1-\alpha),2(1-\alpha);7,4(1-\alpha),4(1-\alpha))}$.

(b) Considering θ tends to ∞, $\delta_{ij5} = 1, \delta_{ij6} = 6 \forall i, j$ and $\gamma_{k1} = \gamma_{k2} = 1; k = 1, 2, 4$, the obtained values of basic variables in a more-for-less solution are $\tilde{x}_{13} =_{Mag} (5, 1(1-\alpha), 1(1-\alpha); 5, 6(1-\alpha), 6(1-\alpha))$,

$$\tilde{x}_{21} \to \infty,$$

$$\tilde{x}_{22} =_{Mag} (15, 1(1-\alpha), 1(1-\alpha); 15, 6(1-\alpha), 6(1-\alpha)),$$

$$\tilde{s}_1 \to \infty,$$

$$\tilde{s}_2 \to \infty,$$

$$\tilde{s}_4 =_{Mag} (1, 1(1-\alpha), 1(1-\alpha); 1, 1(1-\alpha), 1(1-\alpha)),$$

$s_{22}^2 \to \infty$ and the corresponding symmetric triangular intuitionistic fuzzy transportation cost is $\tilde{Z} =_{Mag} \frac{(6,1(1-\alpha),1(1-\alpha);6,6(1-\alpha),6(1-\alpha))}{(7,2(1-\alpha),2(1-\alpha);7,6(1-\alpha),6(1-\alpha))}$.

5.9 Results and Discussion

The results of the existing crisp linear fractional transportation problems with mixed constraints [4] represented by Tables 5.4 and 5.5, obtained by the existing method [4] and the proposed Mehar method-IV, are shown in Table 5.7.

It is obvious from Table 5.7 that

(i) On solving the first existing problem by Gupta et al.'s method [4] as well as by the proposed Mehar method-IV, the same optimal solution and hence, same optimal transportation cost is obtained. While, on solving the second existing problem, different optimal solution and hence, different optimal transportation cost is obtained by the proposed Mehar method-IV and Gupta et al.'s method [4].

(ii) For the first existing problem, different best more-for-less solution is obtained by Gupta et al.'s method [4] and the proposed Mehar method-IV. However, for the second existing problem, no more-for-less solution is obtained on applying Gupta et al.'s method [4] as well as the proposed Mehar method-IV.

Although, it is obvious that the results, obtained by the proposed Mehar method-IV, are better than the existing results [4]. However, the following questions naturally arise in mind.

First question: Why Gupta et al.'s method [4] fails to find the best more-for-less solution for the first problem?

Second question: Why Gupta et al.'s method [4] fails to find an optimal solution for the second problem?

Table 5.7 Results of the existing problems

	First problem [4]		Second problem [4]	
	Gupta et al.'s method [4]	Proposed Mehar method-IV	Gupta et al.'s method [4]	Proposed Mehar method-IV
Optimal solution and corresponding transportation cost	$x_{11}^2 = 0, x_{12}^2 = 0,$ $x_{13}^2 = 5, x_{21}^2 = 8,$ $x_{22}^2 = 15,$ $x_{23}^2 = 0, x_{31}^2 = 0,$ $x_{32}^2 = 0, x_{33}^2 = 0$ and corresponding transportation cost is $\frac{133}{136} = 0.9779$	$x_{11}^2 = \lim_{M\to\infty} \frac{y_{11}^2}{t} = 0,$ $x_{12}^2 = \lim_{M\to\infty} \frac{y_{12}^2}{t} = 0,$ $x_{13}^2 = \lim_{M\to\infty} \frac{y_{13}^2}{t} = 5,$ $x_{21}^2 = \lim_{M\to\infty} \frac{y_{21}^2}{t} = 8,$ $x_{22}^2 = \lim_{M\to\infty} \frac{y_{22}^2}{t} = 15,$ $x_{23}^2 = \lim_{M\to\infty} \frac{y_{23}^2}{t} = 0,$ $x_{31}^2 = \lim_{M\to\infty} \frac{y_{31}^2}{t} = 0,$ $x_{32}^2 = \lim_{M\to\infty} \frac{y_{32}^2}{t} = 0,$ $x_{33}^2 = \lim_{M\to\infty} \frac{y_{33}^2}{t} = 0$ and corresponding transportation cost is $\frac{133}{136} = 0.9779$	$x_{11} = 8,$ $x_{12} = 15,$ $x_{13} = 5,$ $x_{21} = 10,$ $x_{22} = 0,$ $x_{23} = 0,$ $x_{31} = 0,$ $x_{32} = 0,$ $x_{33} = 0$ and corresponding transportation cost is $\frac{170}{183} = 0.9290$	$x_{11}^2 = \lim_{M\to\infty} \frac{y_{11}^2}{t} \to \infty,$ $x_{12}^2 = \lim_{M\to\infty} \frac{y_{12}^2}{t} = 15,$ $x_{13}^2 = \lim_{M\to\infty} \frac{y_{13}^2}{t} = 6,$ $x_{21}^2 = \lim_{M\to\infty} \frac{y_{21}^2}{t} = 10,$ $x_{22}^2 = \lim_{M\to\infty} \frac{y_{22}^2}{t} = 0,$ $x_{23}^2 = \lim_{M\to\infty} \frac{y_{23}^2}{t} = 0,$ $x_{31}^2 = \lim_{M\to\infty} \frac{y_{31}^2}{t} = 0,$ $x_{32}^2 = \lim_{M\to\infty} \frac{y_{32}^2}{t} = 0,$ $x_{33}^2 = \lim_{M\to\infty} \frac{y_{33}^2}{t} = 0$ and corresponding transportation cost is $\frac{5}{6} = 0.8333$

(continued)

Table 5.7 (continued)

	First problem [4]		Second problem [4]	
	Gupta et al.'s method [4]	Proposed Mehar method-IV	Gupta et al.'s method [4]	Proposed Mehar method-IV
Best more-for-less solution and corresponding transportation cost	$x_{11}^2 = 0, x_{12}^2 = 0,$ $x_{13}^2 = 6, x_{21}^2 = 9,$ $x_{22}^2 = 15,$ $x_{23}^2 = 0, x_{31}^2 = 0,$ $x_{32}^2 = 0, x_{33}^2 = 0$ and the corresponding cost is $\frac{141}{147} = 0.9592$	$x_{11}^2 = 0,$ $x_{12}^2 = 0,$ $x_{13}^2 \to \infty,$ $x_{21}^2 = 8,$ $x_{22}^2 = 15,$ $x_{23}^2 = 0,$ $x_{31}^2 = 0,$ $x_{32}^2 = 0,$ $x_{33}^2 = 0$ and corresponding transportation cost is $\frac{2}{4} = 0.5$	No more-for-less solution	No more-for-less solution

5.9.1 *Response of the First Question*

According to the literature [1], Dinkelbach's algorithm [3] may fail to find an optimal solution of crisp linear fractional programming problems having unbounded feasible region. Hence, Dinkelbach's algorithm [3] to transform a crisp linear fractional programming problem into its equivalent crisp linear programming problem can be used only if the feasible region of the considered crisp linear fractional programming problem is closed and bounded. Since, the feasible region of the crisp linear fractional programming problems corresponding to the crisp linear fractional transportation problems with mixed constraints, considered in the first existing problem and the second existing problem, is unbounded. So, Dinkelbach's algorithm [3] cannot be used:

(i) To transform the crisp linear fractional transportation problem with mixed constraints, considered in the first existing problem, into its equivalent crisp linear transportation problem with mixed constraints.
(ii) To transform the crisp linear fractional transportation problem with mixed constraints, considered in the second existing problem, into its equivalent crisp linear transportation problem with mixed constraints.

However, it can be easily verified that to obtain the results, shown in Table 5.7, Gupta et al. [4] have used Dinkelbach's algorithm [3] to transform

(i) The crisp linear fractional transportation problem with mixed constraints, considered in the first existing problem, into a crisp linear transportation problem with mixed constraints.

(ii) The crisp linear fractional transportation problem with mixed constraints, considered in the second existing problem, into a crisp linear transportation problem with mixed constraints.

The problem (*P*5.18) is considered to validate the claim that Dinkelbach's algorithm [3] fails to find an optimal solution of crisp linear fractional programming problems having unbounded feasible region.

Problem (***P***5.18) *Maximize* $\left(\frac{2x_1+3x_2+6}{4x_1+5x_2+18}\right)$
Subject to

$$x_1 + x_2 \geq 1,$$

$$x_i \geq 0, i = 1, 2.$$

It is obvious that the feasible region of the crisp linear fractional programming problem (*P*5.18) is unbounded. Also, it is obvious that $x_1 = 0$ and $x_2 \to \infty$ is an optimal solution of the crisp linear fractional programming problem (*P*5.18) and the corresponding optimal value is $\lim\limits_{x_2\to\infty}\left(\frac{2x_1+3x_2+6}{4x_1+5x_2+18}\right) = \lim\limits_{x_2\to\infty}\left(\frac{2(0)+3x_2+6}{4(0)+5x_2+18}\right) = \frac{3}{5} = 0.6.$

While, the following clearly indicates that on applying Dinkelbach's algorithm [3], the obtained optimal solution of the crisp linear fractional programming problem (*P*5.18) is $x_1 = 1, x_2 = 0$ and the corresponding optimal value is $\frac{8}{22} = 0.363636$. This clearly indicates that Dinkelbach's algorithm [3] may fail to find an optimal solution of the crisp linear fractional programming problem having unbounded feasible region.

Since $x_1 = 1$ and $x_2 = 0$ is a feasible solution of the crisp linear fractional programming problem (*P*5.18). Therefore, according to Dinkelbach's algorithm [3], using $\delta = \frac{2(1)+3(0)+6}{4(1)+5(0)+18} = \frac{8}{22}$, the crisp linear fractional programming problem (*P*5.18) can be transformed into its equivalent crisp linear programming problem (*P*5.19).

Problem (***P***5.19) *Maximize* $\left(2x_1 + 3x_2 + 6 - \frac{8}{22}(4x_1 + 5x_2 + 18)\right)$
Subject to

$$x_1 + x_2 \geq 1,$$

$$x_i \geq 0, i = 1, 2.$$

It can be easily verified that the value of the objective function of the crisp linear programming problem (*P*5.19) is zero. So, according to Dinkelbach's algorithm [3], $x_1 = 1$ and $x_2 = 0$ is an optimal solution of the crisp linear fractional programming problem (*P*5.18).

Table 5.8 Crisp balanced linear fractional transportation problem

	D_1	D_2	...	D_n	Availability
S_1	c_{11} d_{11}	c_{12} d_{12}	...	c_{1n} d_{1n}	$= a_1$
S_2	c_{21} d_{21}	c_{22} d_{22}	...	c_{2n} d_{2n}	$= a_2$
$\vdots$	$\vdots$	$\vdots$	$\vdots$	$\vdots$	$\vdots$
S_m	c_{m1} d_{m1}	c_{m2} d_{m2}	...	c_{mn} d_{mn}	$= a_m$
Demand	$= b_1$	$= b_2$	...	$= b_n$	$\sum_{i=1}^{m} a_i = \sum_{j=1}^{n} b_j$

5.9.2 Response of the Second Question

Using the crisp linear fractional programming problem ($P5.20$) of a balanced crisp linear fractional transportation problem, Verma and Puri [6] proved that the condition (5.1) can be used to check that at least one more-for-less solution will exist or not for the balanced crisp linear fractional transportation problem represented by Table 5.8.

$$\left(\sum_{i=1}^{m}\sum_{j=1}^{n} d_{ij}\left(x_{ij}^{*}\right)^{B}\right)\left(u_p + v_q\right) - \left(\sum_{i=1}^{m}\sum_{j=1}^{n} c_{ij}\left(x_{ij}^{*}\right)^{B}\right)\left(u_p^{'} + v_q^{'}\right) < 0 \quad (5.1)$$

where u_i, v_j, $u_i^{'}$ and $v_j^{'}$ are obtained by solving the system of equations $c_{ij} = u_i + v_j$ and $d_{ij} = u_i^{'} + v_j^{'}$ corresponding to optimal basic decision variable $\left(x_{ij}^{*}\right)^{B}$.

Problem ($P5.20$) *Minimize* $\left(Z = \frac{\sum_{i=1}^{m}\sum_{j=1}^{n} c_{ij}x_{ij}}{\sum_{i=1}^{m}\sum_{j=1}^{n} d_{ij}x_{ij}}\right)$

Subject to

$$\sum_{j=1}^{n} x_{ij} = a_i;\ i = 1, 2, \ldots, m;$$

$$\sum_{i=1}^{m} x_{ij} = b_j;\ j = 1, 2, \ldots, n;$$

$$x_{ij} \geq 0;\ i = 1, 2, \ldots, m;\ j = 1, 2, \ldots, n.$$

It can be easily verified that Gupta et al. [4] have also used the condition (5.1) to check that at least one more-for-less solution of the crisp linear fractional transportation problem with mixed constraints, represented by Table 5.1, will exist or not. While, in this section, it is shown that the condition (5.1) cannot be used to check the existence of at least one more-for-less solution of crisp linear fractional transportation problem with mixed constraints.

5.9.2.1 Origin of the Condition

Gupta et al. [4] have obtained the condition (5.1) as follows.

Let $x^* = \left\{x_{ij}^*; i \in I, j \in J\right\}$ be an optimal solution of the crisp linear fractional programming problem ($P5.1$). Then, $Z(x^*) = \frac{\sum_{i=1}^{m}\sum_{j=1}^{n} c_{ij}x_{ij}^*}{\sum_{i=1}^{m}\sum_{j=1}^{n} d_{ij}x_{ij}^*}$ can be re-written as $Z(x^*) = \frac{\sum_{i=1}^{m}\sum_{j=1}^{n} c_{ij}\left(x_{ij}^*\right)^{NB} + \sum_{i=1}^{m}\sum_{j=1}^{n} c_{ij}\left(x_{ij}^*\right)^{B}}{\sum_{i=1}^{m}\sum_{j=1}^{n} d_{ij}\left(x_{ij}^*\right)^{NB} + \sum_{i=1}^{m}\sum_{j=1}^{n} d_{ij}\left(x_{ij}^*\right)^{B}}$, where $\left(x_{ij}^*\right)^{NB}$ represents an optimal non-basic decision variable and $\left(x_{ij}^*\right)^{B}$ represents an optimal basic decision variable.

Since, for the optimal solution of the problem ($P5.1$), the following conditions will be satisfied.

(i) The value of all the optimal non-basic decision variables i.e., $\left(x_{ij}^*\right)^{NB}$ will be zero.
(ii) For an optimal basic decision variable $\left(x_{ij}^*\right)^{B}$, $c_{ij} = u_i + v_j$ and $d_{ij} = u_i^{'} + v_j^{'}$.

So,

$$
\begin{aligned}
Z(x^*) &= \frac{\sum_{i=1}^{m}\sum_{j=1}^{n} c_{ij}\left(x_{ij}^*\right)^{NB} + \sum_{i=1}^{m}\sum_{j=1}^{n} c_{ij}\left(x_{ij}^*\right)^{B}}{\sum_{i=1}^{m}\sum_{j=1}^{n} d_{ij}\left(x_{ij}^*\right)^{NB} + \sum_{i=1}^{m}\sum_{j=1}^{n} d_{ij}\left(x_{ij}^*\right)^{B}} \\
&= \frac{\sum_{i=1}^{m}\sum_{j=1}^{n}\left(u_i + v_j\right)\left(x_{ij}^*\right)^{B}}{\sum_{i=1}^{m}\sum_{j=1}^{n}\left(u_i^{'} + v_j^{'}\right)\left(x_{ij}^*\right)^{B}} \\
&= \frac{\sum_{i=1}^{m}\sum_{j=1}^{n} u_i\left(x_{ij}^*\right)^{B} + \sum_{i=1}^{m}\sum_{j=1}^{n} v_j\left(x_{ij}^*\right)^{B}}{\sum_{i=1}^{m}\sum_{j=1}^{n} u_i^{'}\left(x_{ij}^*\right)^{B} + \sum_{i=1}^{m}\sum_{j=1}^{n} v_j^{'}\left(x_{ij}^*\right)^{B}} \\
&= \frac{\sum_{i=1}^{m} u_i\left(\sum_{j=1}^{n}\left(x_{ij}^*\right)^{B}\right) + \sum_{j=1}^{n} v_j\left(\sum_{i=1}^{m}\left(x_{ij}^*\right)^{B}\right)}{\sum_{i=1}^{m} u_i^{'}\left(\sum_{j=1}^{n}\left(x_{ij}^*\right)^{B}\right) + \sum_{j=1}^{n} v_j^{'}\left(\sum_{i=1}^{m}\left(x_{ij}^*\right)^{B}\right)}
\end{aligned} \tag{5.2}
$$

Assuming $\sum_{j=1}^{n} x_{ij} = a_i; i = 1, 2, \ldots, m$ and $\sum_{i=1}^{m} x_{ij} = b_j; j = 1, 2, \ldots, n$.

$$
Z(x^*) = \frac{\sum_{i=1}^{m} u_i a_i + \sum_{j=1}^{n} v_j b_j}{\sum_{i=1}^{m} u_i^{'} a_i + \sum_{j=1}^{n} v_j^{'} b_j} \tag{5.3}
$$

If the availability of the pth source and the demand of the qth destination is increased by the positive quantity (say, λ). Then, the total transportation cost for the modified problem will be

$$Z'(x^*) = \frac{\sum_{i=1}^{m} u_i a_i + \sum_{j=1}^{n} v_j b_j + \lambda(u_p + v_q)}{\sum_{i=1}^{m} u'_i a_i + \sum_{j=1}^{n} v'_j b_j + \lambda(u'_p + v'_q)}$$

It is obvious that.

$$Z'(x^*) - Z(x^*) = \frac{\lambda\left(\left(\sum_{i=1}^{m}\sum_{j=1}^{n} d_{ij}\left(x_{ij}^*\right)^B\right)(u_p + v_q) - \left(\sum_{i=1}^{m}\sum_{j=1}^{n} c_{ij}\left(x_{ij}^*\right)^B\right)\left(u'_p + v'_q\right)\right)}{\left(\sum_{i=1}^{m}\sum_{j=1}^{n} d_{ij}\left(x_{ij}^*\right)^B\right)\left(\left(\sum_{i=1}^{m}\sum_{j=1}^{n} d_{ij}\left(x_{ij}^*\right)^B\right) + \theta\left(u'_p + v'_q\right)\right)}.$$

Since, λ is a positive real number. So, $Z'(x^*)$ will be less than $Z(x^*)$ if $\left(\sum_{i=1}^{m}\sum_{j=1}^{n} d_{ij}\left(x_{ij}^*\right)^B\right)(u_p + v_q) - \left(\sum_{i=1}^{m}\sum_{j=1}^{n} c_{ij}\left(x_{ij}^*\right)^B\right)\left(u'_p + v'_q\right) < 0$. Therefore, according to Gupta et al. [4], a more-for-less solution for the crisp linear fractional transportation problem with mixed constraints, represented by Table 5.1, will exist only if there exist a $\lambda > 0$ and variables u_p, v_q, u'_p and v'_q such that $\left(\sum_{i=1}^{m}\sum_{j=1}^{n} d_{ij}\left(x_{ij}^*\right)^B\right)(u_p + v_q) - \left(\sum_{i=1}^{m}\sum_{j=1}^{n} c_{ij}\left(x_{ij}^*\right)^B\right)\left(u'_p + v'_q\right) < 0$.

5.9.2.2 Validity of the Claim

It is obvious from Sect. 5.9.2.1 that the condition (5.1) is obtained with the help of the expression (5.3). Also, it is obvious from Sect. 5.9.2.1 that the expression (5.3) is obtained from the expression (5.2) by considering the following assumption:

$$\sum_{j=1}^{n} x_{ij} = a_i;\ i = 1, 2, \ldots, m \quad \text{and} \quad \sum_{i=1}^{m} x_{ij} = b_j;\ j = 1, 2, \ldots, n$$

i.e., the sign of each row and each column of Table 5.1 is exactly equal. While, this assumption is not valid in case of a crisp linear fractional transportation problem with mixed constraints as the sign of at least one row/column of Table 5.1 will not exactly equal. Hence, the condition (5.1) cannot be used to check the existence of at least one more-for-less solution of crisp linear fractional transportation problem with mixed constraints.

5.10 Conclusions

In this chapter, the sufficient condition-IV is proposed to check the existence of at least one more-for-less solution of symmetric triangular intuitionistic fuzzy linear fractional transportation problems with mixed constraints. Also, a method is proposed to find all more-for-less solutions (if exist) of symmetric triangular intuitionistic fuzzy linear fractional transportation problems with mixed constraints. Furthermore, it is pointed out that an optimal solution and/or more-for-less solution, obtained by the existing method [4], may or may not be correct. While, all more-for-less solutions, obtained by the proposed Mehar method-IV, will always be correct.

References

1. E.B. Bajalinov, *Linear-Fractional Programming Theory, Methods, Applications and Software* (Kluwer Academic Publishers, Boston, 2003)
2. A. Charnes, W.W. Cooper, Programming with linear fractional functionals. Naval. Res. Logistics Q. **9**, 181–186 (1962)
3. W. Dinkelbach, On nonlinear fractional programming. Manage. Sci. **13**, 492–498 (1967)
4. A. Gupta, S. Khanna, M.C. Puri, A paradox in linear fractional transportation problem with mixed constraints. Optimization **27**, 375–387 (1993)
5. H.A. Taha, *Operations Research: an Introduction* (Pearson Prentice Hall Upper Saddle River, N.J., 2013)
6. V. Verma, M.C. Puri, On a paradox in linear fractional transportation problem, in *Recent Developments in Mathematical Programming, Published on Behalf of Australian Society for Operations Research* ed. by S. Kumar (Gordan and Breach Science Publishers, 1991), pp. 413–424

Chapter 6
Some Open Research Problems

In this chapter, some open research problems are discussed.

6.1 First Open Research Problem

It is pertinent to mention that in the Mehar method-I and Mehar method-II, proposed in Chaps. 2 and 3, the relation $\text{Mag}\left(*_{i=1}^{n}\tilde{A}_i\right) = *_{i=1}^{n}\text{Mag}\left(\tilde{A}_i\right)$, where $*$ is $\oplus$ or $\otimes$ is used. Also, it is pertinent to mention that if $\tilde{A}_i$ is a symmetric triangular fuzzy number. Then, this relation will necessarily hold. However, if $\tilde{A}_i$ is a non-symmetric triangular fuzzy number. Then, this relation will not necessarily hold. Hence, the proposed Mehar method-I and the proposed Mehar method-II cannot be used

(i) To find more-for-less solutions of such fuzzy balanced transportation problems in which each parameter is not represented by a symmetric triangular fuzzy number.
(ii) To find more-for-less solutions of such fuzzy transportation problems with mixed constraints in which each parameter is not represented by a symmetric triangular fuzzy number.

The following clearly indicates that if $\tilde{A}_i$ is a non-symmetric triangular fuzzy number. Then, the relation $\text{Mag}\left(*_{i=1}^{n}\tilde{A}_i\right) = *_{i=1}^{n}\text{Mag}\left(\tilde{A}_i\right)$, where $*$ is $\oplus$ or $\otimes$, used in the proposed Mehar method-I and the proposed Mehar method-II, will not necessarily hold.

If $\tilde{A}_1 = \left(a_1^2, \delta_1(1-\alpha), \gamma_1(1-\alpha)\right)$ and $\tilde{A}_2 = \left(a_2^2, \delta_2(1-\alpha), \gamma_2(1-\alpha)\right)$, where $\alpha \in (0, 1]$, $\delta_1 = a_1^2 - a_1^1$, $\gamma_1 = a_1^3 - a_1^2$, $\delta_2 = a_2^2 - a_2^1$, $\gamma_2 = a_2^3 - a_2^2$ are two triangular fuzzy numbers.

Then, according to the existing arithmetic operation [2],

T. K. Bhatia et al., *More-for-Less Solutions in Fuzzy Transportation Problems*, Studies in Fuzziness and Soft Computing 426,
https://doi.org/10.1007/978-3-031-30337-1_6

$$\tilde{A}_1 * \tilde{A}_2 = (a_1^2 * a_2^2, \text{maximum}\{\delta_1(1-\alpha), \delta_2(1-\alpha)\},) \\ \text{maximum}\{\gamma_1(1-\alpha), \gamma_2(1-\alpha)\}) \text{ where} * \text{is} \oplus \text{ or } \otimes.$$

Using the existing expression [2], $\text{Mag}\left(\tilde{A}_i\right) = \frac{\gamma_i(1-\alpha)+4a_i^2-\delta_i(1-\alpha)}{4}; i = 1, 2,$

$$\begin{aligned}\text{Mag}\left(\tilde{A}_1 * \tilde{A}_2\right) &= \text{Mag}\left(a_1^2 * a_2^2, \text{maximum}\{\delta_1(1-\alpha), \delta_2(1-\alpha)\},\right. \\ &\quad \left.\text{maximum}\{\gamma_1(1-\alpha), \gamma_2(1-\alpha)\}\right) \\ &= \frac{\text{maximum}\{\gamma_1(1-\alpha), \gamma_2(1-\alpha)\} + 4\left(a_1^2 * a_2^2\right) - \text{maximum}\{\delta_1(1-\alpha), \delta_2(1-\alpha)\}}{4}\end{aligned} \quad (6.1)$$

$$\text{Mag}\left(\tilde{A}_1\right) = \frac{\gamma_1(1-\alpha) + 4a_1^2 - \delta_1(1-\alpha)}{4} \quad (6.2)$$

$$\text{Mag}\left(\tilde{A}_2\right) = \frac{\gamma_2(1-\alpha) + 4a_2^2 - \delta_2(1-\alpha)}{4} \quad (6.3)$$

From (6.2) and (6.3),

$$\begin{aligned}&\text{Mag}\left(\tilde{A}_1\right) * \text{Mag}\left(\tilde{A}_2\right) \\ &= \frac{\gamma_1(1-\alpha) + 4a_1^2 - \delta_1(1-\alpha)}{4} * \frac{\gamma_2(1-\alpha) + 4a_2^2 - \delta_2(1-\alpha)}{4}\end{aligned} \quad (6.4)$$

It is obvious from (6.1) and (6.4) that in general $\text{Mag}\left(\tilde{A}_1 * \tilde{A}_2\right) \neq \text{Mag}\left(\tilde{A}_1\right) * \text{Mag}\left(\tilde{A}_2\right)$.

In future, one may generalize the proposed Mehar method-I and the proposed Mehar method-II/propose new methods to overcome the above-mentioned limitation of the proposed Mehar method-I and the proposed Mehar method-II respectively.

6.2 Second Open Research Problem

It is pertinent to mention that in the Mehar method-III, proposed in Chap. 4, the relation $\text{Mag}\left(*_{i=1}^{n}\tilde{A}_i\right) = *_{i=1}^{n}\text{Mag}\left(\tilde{A}_i\right)$, where $*$ is $\oplus$ or $\otimes$ is used. Also, it is pertinent to mention that if $\tilde{A}_i$ is a symmetric triangular intuitionistic fuzzy number. Then, this relation will necessarily hold. However, if $\tilde{A}_i$ is a non-symmetric triangular intuitionistic fuzzy number. Then, this relation will not necessarily hold. Hence, the proposed Mehar method-III cannot be used to find more-for-less solutions of such intuitionistic fuzzy transportation problems with mixed constraints in which each parameter is not represented by a symmetric triangular intuitionistic fuzzy number.

The following clearly indicates that if $\tilde{A}_i$ is a non-symmetric triangular intuitionistic fuzzy number. Then, the relation $\text{Mag}\left(*_{i=1}^{n}\tilde{A}_i\right) = *_{i=1}^{n}\text{Mag}\left(\tilde{A}_i\right)$, where $*$ is $\oplus$ or $\otimes$, used in the proposed Mehar method-III, will not necessarily hold.

If $\tilde{A}_1 = \left(a_1^2, \delta_1(1-\alpha), \gamma_1(1-\alpha); a_1^2, \delta_1^{'}(1-\alpha), \gamma_1^{'}(1-\alpha)\right)$ and $\tilde{A}_2 = \left(a_2^2, \delta_2(1-\alpha), \gamma_2(1-\alpha); a_2^2, \delta_2^{'}(1-\alpha), \gamma_2^{'}(1-\alpha)\right)$, where $\alpha \in (0,1]$, $\delta_1 = a_1^2 - a_1^1$, $\gamma_1 = a_1^3 - a_1^2$, $\delta_1^{'} = a_1^2 - A_1^1$, $\gamma_1^{'} = A_1^3 - a_1^2$, $\delta_2 = a_2^2 - a_2^1$, $\gamma_2 = a_2^3 - a_2^2$, $\delta_2^{'} = a_2^2 - A_2^1$, $\gamma_2^{'} = A_2^3 - a_2^2$ are two triangular intuitionistic fuzzy numbers.

Then, according to the extended arithmetic operation, discussed in Chap. 4,

$$\begin{aligned}\tilde{A}_1 * \tilde{A}_2 =& \big(a_1^2 * a_2^2, \text{maximum}\{\delta_1(1-\alpha), \delta_2(1-\alpha)\},\\ &\text{maximum}\{\gamma_1(1-\alpha), \gamma_2(1-\alpha)\};\\ &a_1^2 * a_2^2, \text{maximum}\left\{\delta_1^{'}(1-\alpha), \delta_2^{'}(1-\alpha)\right\},\\ &\text{maximum}\left\{\gamma_1^{'}(1-\alpha), \gamma_2^{'}(1-\alpha)\right\}\big) \quad \text{where } * \text{ is } \oplus \text{ or } \otimes.\end{aligned}$$

Using the expression (4.4), discussed in Sect. 4.3 of Chap. 4,

$$\begin{aligned}\text{Mag}\left(\tilde{A}_1 * \tilde{A}_2\right) =& \text{Mag}\big(a_1^2 * a_2^2, \text{maximum}\{\delta_1(1-\alpha), \delta_2(1-\alpha)\},\\ &\text{maximum}\{\gamma_1(1-\alpha), \gamma_2(1-\alpha)\};\\ &a_1^2 * a_2^2, \text{maximum}\left\{\delta_1^{'}(1-\alpha), \delta_2^{'}(1-\alpha)\right\},\\ &\text{maximum}\left\{\gamma_1^{'}(1-\alpha), \gamma_2^{'}(1-\alpha)\right\}\big)\\ =& \frac{1}{8}\big(\text{maximum}\{\gamma_1(1-\alpha), \gamma_2(1-\alpha)\} + 8\left(a_1^2 * a_2^2\right)\\ &- \text{maximum}\{\delta_1(1-\alpha), \delta_2(1-\alpha)\}\\ &+ \max imum\left\{\gamma_1^{'}(1-\alpha), \gamma_2^{'}(1-\alpha)\right\}\\ &- \text{maximum}\left\{\delta_1^{'}(1-\alpha), \delta_2^{'}(1-\alpha)\right\}\big)\end{aligned} \tag{6.5}$$

$$\text{Mag}\left(\tilde{A}_1\right) = \frac{\gamma_1(1-\alpha) + 8a_1^2 - \delta_1(1-\alpha) + \gamma_1^{'}(1-\alpha) - \delta_1^{'}(1-\alpha)}{8} \tag{6.6}$$

$$\text{Mag}\left(\tilde{A}_2\right) = \frac{\gamma_2(1-\alpha) + 8a_2^2 - \delta_2(1-\alpha) + \gamma_2^{'}(1-\alpha) - \delta_2^{'}(1-\alpha)}{8} \tag{6.7}$$

From (6.6) and (6.7),

$$\begin{aligned}\text{Mag}\left(\tilde{A}_1\right) * \text{Mag}\left(\tilde{A}_2\right) =& \frac{\gamma_1(1-\alpha) + 8a_1^2 - \delta_1(1-\alpha) + \gamma_1^{'}(1-\alpha) - \delta_1^{'}(1-\alpha)}{8}\\ &* \frac{\gamma_2(1-\alpha) + 8a_2^2 - \delta_2(1-\alpha) + \gamma_2^{'}(1-\alpha) - \delta_2^{'}(1-\alpha)}{8}\end{aligned} \tag{6.8}$$

It is obvious from (6.5) and (6.8) that in general $\text{Mag}\left(\tilde{A}_1 * \tilde{A}_2\right) \neq \text{Mag}\left(\tilde{A}_1\right) * \text{Mag}\left(\tilde{A}_2\right)$.

In future, one may generalize the proposed Mehar method-III/propose a new method to overcome the above-mentioned limitation of the proposed Mehar method-III.

6.3 Third Open Research Problem

It is pertinent to mention that in the Mehar method-IV, proposed in Chap. 4, the relations the relations $\text{Mag}\left(*_{i=1}^{n}\tilde{A}_i\right) = *_{i=1}^{n}\text{Mag}\left(\tilde{A}_i\right)$, where $*$ is $\oplus$ or $\otimes$ and $\text{Mag}\left(\frac{\tilde{A}_1}{\tilde{A}_2}\right) = \frac{\text{Mag}(\tilde{A}_1)}{\text{Mag}(\tilde{A}_2)}$ are used. Also, it is pertinent to mention that if $\tilde{A}_i$ is a symmetric triangular intuitionistic fuzzy number. Then, these relations will necessarily hold. However, if $\tilde{A}_i$ is a non-symmetric triangular intuitionistic fuzzy number. Then, these relations will not necessarily hold. Hence, the proposed Mehar method-IV cannot be used to find more-for-less solutions of such intuitionistic fuzzy linear fractional transportation problems with mixed constraints in which each parameter is not represented by a symmetric triangular intuitionistic fuzzy number.

The following clearly indicates that if $\tilde{A}_i$ is a non-symmetric triangular intuitionistic fuzzy number. Then, the relation $\text{Mag}\left(\frac{\tilde{A}_1}{\tilde{A}_2}\right) = \frac{\text{Mag}(\tilde{A}_1)}{\text{Mag}(\tilde{A}_2)}$, used in the proposed Mehar method-IV, will not necessarily hold.

If $\tilde{A}_1 = \left(a_1^2, \delta_1(1-\alpha), \gamma_1(1-\alpha); a_1^2, \delta_1^{'}(1-\alpha), \gamma_1^{'}(1-\alpha)\right)$ and $\tilde{A}_2 = \left(a_2^2, \delta_2(1-\alpha), \gamma_2(1-\alpha); a_2^2, \delta_2^{'}(1-\alpha), \gamma_2^{'}(1-\alpha)\right); a_2^2 \neq 0$, where $\alpha \in (0, 1]$, $\delta_1 = a_1^2 - a_1^1$, $\gamma_1 = a_1^3 - a_1^2$, $\delta_1^{'} = a_1^2 - A_1^1$, $\gamma_1^{'} = A_1^3 - a_1^2$, $\delta_2 = a_2^2 - a_2^1$, $\gamma_2 = a_2^3 - a_2^2$, $\delta_2^{'} = a_2^2 - A_2^1$, $\gamma_2^{'} = A_2^3 - a_2^2$ are two triangular intuitionistic fuzzy numbers.

Then, according to the extended arithmetic operation, discussed in Chap. 4,

$$\begin{aligned}\frac{\tilde{A}_1}{\tilde{A}_2} = &\left(\frac{a_1^2}{a_2^2}, \text{maximum}\{\delta_1(1-\alpha), \delta_2(1-\alpha)\}, \text{maximum}\{\gamma_1(1-\alpha), \gamma_2(1-\alpha)\};\right.\\ &\left.\frac{a_1^2}{a_2^2}, \text{maximum}\left\{\delta_1^{'}(1-\alpha), \delta_2^{'}(1-\alpha)\right\}, \text{maximum}\left\{\gamma_1^{'}(1-\alpha), \gamma_2^{'}(1-\alpha)\right\}\right).\end{aligned}$$

Using the expression (4.4), proposed in Sect. 4.3 of Chap. 4,

$$\begin{aligned}\text{Mag}\left(\frac{\tilde{A}_1}{\tilde{A}_2}\right) = &\text{Mag}\left(\frac{a_1^2}{a_2^2}, \text{maximum}\{\delta_1(1-\alpha), \delta_2(1-\alpha)\},\right.\\ &\text{maximum}\{\gamma_1(1-\alpha), \gamma_2(1-\alpha)\};\end{aligned}$$

$$\frac{a_1^2}{a_2^2}, \text{maximum}\{\delta_1'(1-\alpha), \delta_2'(1-\alpha)\},$$
$$\text{maximum}\{\gamma_1'(1-\alpha), \gamma_2'(1-\alpha)\})$$
$$= \frac{1}{8}\left(\text{maximum}\{\gamma_1(1-\alpha), \gamma_2(1-\alpha)\} + 8\left(\frac{a_1^2}{a_2^2}\right)\right.$$
$$- \text{maximum}\{\delta_1(1-\alpha), \delta_2(1-\alpha)\}$$
$$+ \text{maximum}\{\gamma_1'(1-\alpha), \gamma_2'(1-\alpha)\}$$
$$\left.-\text{maximum}\{\delta_1'(1-\alpha), \delta_2'(1-\alpha)\}\right) \tag{6.9}$$

$$\text{Mag}\left(\tilde{A}_1\right) = \frac{\gamma_1(1-\alpha) + 8a_1^2 - \delta_1(1-\alpha) + \gamma_1'(1-\alpha) - \delta_1'(1-\alpha)}{8} \tag{6.10}$$

$$\text{Mag}\left(\tilde{A}_2\right) = \frac{\gamma_2(1-\alpha) + 8a_2^2 - \delta_2(1-\alpha) + \gamma_2'(1-\alpha) - \delta_2'(1-\alpha)}{8} \tag{6.11}$$

From (6.10) and (6.11),

$$\frac{\text{Mag}\left(\tilde{A}_1\right)}{\text{Mag}\left(\tilde{A}_2\right)} = \frac{\gamma_1(1-\alpha) + 8a_1^2 - \delta_1(1-\alpha) + \gamma_1'(1-\alpha) - \delta_1'(1-\alpha)}{\gamma_2(1-\alpha) + 8a_2^2 - \delta_2(1-\alpha) + \gamma_2'(1-\alpha) - \delta_2'(1-\alpha)} \tag{6.12}$$

It is obvious from (6.9) and (6.12) that in general $\text{Mag}\left(\frac{\tilde{A}_1}{\tilde{A}_2}\right) \neq \frac{\text{Mag}(\tilde{A}_1)}{\text{Mag}(\tilde{A}_2)}$.

In future, one may generalize the proposed Mehar method-IV/propose a new method to overcome the above-mentioned limitation of the proposed Mehar method-IV.

6.4 Fourth Open Research Problem

In the Mehar method-I and Mehar method-II, proposed in Chaps. 2 and 3, the following existing arithmetic operations [2] are used.

If $\tilde{A}_1 = \left(a_1^2, \delta_1(1-\alpha), \gamma_1(1-\alpha)\right)$ and $\tilde{A}_2 = \left(a_2^2, \delta_2(1-\alpha), \gamma_2(1-\alpha)\right)$, where $\alpha \in (0, 1]$, $\delta_1 = a_1^2 - a_1^1$, $\gamma_1 = a_1^3 - a_1^2$, $\delta_2 = a_2^2 - a_2^1$, $\gamma_2 = a_2^3 - a_2^2$ are two triangular fuzzy numbers. Then,

$$\tilde{A}_1 * \tilde{A}_2 = (a_1^2 * a_2^2, \text{maximum}\{\delta_1(1-\alpha), \delta_2(1-\alpha)\},$$
$$\text{maximum}\{\gamma_1(1-\alpha), \gamma_2(1-\alpha)\}) \text{ where } * \text{ is } \oplus \text{ or } \otimes .$$

While, in general, the following arithmetic operations [1] are used.

If $\tilde{A}_1 = \left(a_1^1, a_1^2, a_1^3\right)$ and $\tilde{A}_2 = \left(a_2^1, a_2^2, a_2^3\right)$ are two triangular fuzzy numbers. Then,

(i) $\tilde{A}_1 \oplus \tilde{A}_2 = \left(a_1^1 + a_2^1, a_1^2 + a_2^2, a_1^3 + a_2^3\right)$.
(ii) $\tilde{A}_1 \otimes \tilde{A}_2 = \left(\text{minimum}\left\{a_1^1 a_2^1, a_1^1 a_2^3, a_1^3 a_2^1, a_1^3 a_2^3\right\},\right.$
$\left. a_1^2 a_2^2, \text{maximum}\left\{a_1^1 a_2^1, a_1^1 a_2^3, a_1^3 a_2^1, a_1^3 a_2^3\right\}\right.$.

In future, it may be tried to propose a method to find more-for-less solutions of fuzzy transportation problems with mixed constraints based on these arithmetic operations.

6.5 Fifth Open Research Problem

In the Mehar method-III and Mehar method-IV, proposed in Chaps. 4 and 5, the following extended arithmetic operations are used.

If $\tilde{A}_1 = \left(a_1^2, \delta_1(1-\alpha), \gamma_1(1-\alpha); a_1^2, \delta_1^{'}(1-\alpha), \gamma_1^{'}(1-\alpha)\right)$ and $\tilde{A}_2 = \left(a_2^2, \delta_2(1-\alpha), \gamma_2(1-\alpha); a_2^2, \delta_2^{'}(1-\alpha), \gamma_2^{'}(1-\alpha)\right)$, where $\alpha \in (0, 1]$, $\delta_1 = a_1^2 - a_1^1$, $\gamma_1 = a_1^3 - a_1^2$, $\delta_1^{'} = a_1^2 - A_1^1$, $\gamma_1^{'} = A_1^3 - a_1^2$, $\delta_2 = a_2^2 - a_2^1$, $\gamma_2 = a_2^3 - a_2^2$, $\delta_2^{'} = a_2^2 - A_2^1$, $\gamma_2^{'} = A_2^3 - a_2^2$ are two triangular intuitionistic fuzzy numbers. Then,

(i)
$$\begin{aligned} \tilde{A}_1 * \tilde{A}_2 = &\left(a_1^2 * a_2^2, \text{maximum}\{\delta_1(1-\alpha), \delta_2(1-\alpha)\},\right. \\ &\text{maximum}\{\gamma_1(1-\alpha), \gamma_2(1-\alpha)\}; \\ &a_1^2 * a_2^2, \text{maximum}\left\{\delta_1^{'}(1-\alpha), \delta_2^{'}(1-\alpha)\right\}, \\ &\left.\text{maximum}\left\{\gamma_1^{'}(1-\alpha), \gamma_2^{'}(1-\alpha)\right\}\right) \end{aligned}$$
where $*$ is $\oplus$ or $\otimes$.

(ii)
$$\begin{aligned} \frac{\tilde{A}_1}{\tilde{A}_2} = &\left(\frac{a_1^2}{a_2^2}, \text{maximum}\{\delta_1(1-\alpha), \delta_2(1-\alpha)\},\right. \\ &\text{maximum}\{\gamma_1(1-\alpha), \gamma_2(1-\alpha)\}; \\ &\frac{a_1^2}{a_2^2}, \text{maximum}\left\{\delta_1^{'}(1-\alpha), \delta_2^{'}(1-\alpha)\right\}, \\ &\left.\text{maximum}\left\{\gamma_1^{'}(1-\alpha), \gamma_2^{'}(1-\alpha)\right\}\right). \end{aligned}$$

While, in general, the following arithmetic operations [1] are used.

If $\tilde{A}_1 = \left(a_1^1, a_1^2, a_1^3; A_1^1, a_1^2, A_1^3\right)$ and $\tilde{A}_2 = \left(a_2^1, a_2^2, a_2^3; A_2^1, a_2^2, A_2^3\right)$ are two triangular intuitionistic fuzzy numbers. Then,

(i)

$$\tilde{A}_1 \oplus \tilde{A}_2 = \left(a_1^1 + a_2^1, a_1^2 + a_2^2, a_1^3 + a_2^3; A_1^1 + A_2^1, a_1^2 + a_2^2, A_1^3 + A_2^3\right)$$

(ii)

$$\tilde{A}_1 \otimes \tilde{A}_2 = \begin{pmatrix} \text{minimum}\{a_1^1 a_2^1, a_1^1 a_2^3, a_1^3 a_2^1, a_1^3 a_2^3\}, a_1^2 a_2^2, \text{maximum}\{a_1^1 a_2^1, a_1^1 a_2^3, a_1^3 a_2^1, a_1^3 a_2^3\}; \\ \text{minimum}\{A_1^1 a_2^1, A_1^1 a_2^3, A_1^3 a_2^1, A_1^3 a_2^3\}, a_1^2 a_2^2, \text{maximum}\{A_1^1 a_2^1, A_1^1 a_2^3, A_1^3 a_2^1, A_1^3 a_2^3\} \end{pmatrix}$$

(iii)

$$\frac{\tilde{A}_1}{\tilde{A}_2} = \begin{pmatrix} \text{minimum}\left\{\frac{a_1^1}{a_2^1}, \frac{a_1^1}{a_2^3}, \frac{a_1^3}{a_2^1}, \frac{a_1^3}{a_2^3}\right\}, \frac{a_1^2}{a_2^2}, \text{maximum}\left\{\frac{a_1^1}{a_2^1}, \frac{a_1^1}{a_2^3}, \frac{a_1^3}{a_2^1}, \frac{a_1^3}{a_2^3}\right\}; \\ \text{minimum}\left\{\frac{A_1^1}{A_2^1}, \frac{A_1^1}{A_2^3}, \frac{A_1^3}{A_2^1}, \frac{a_1^3}{a_2^3}\right\}, \frac{a_1^2}{a_2^2}, \text{maximum}\left\{\frac{A_1^1}{A_2^1}, \frac{A_1^1}{A_2^3}, \frac{A_1^3}{A_2^1}, \frac{a_1^3}{a_2^3}\right\} \end{pmatrix}$$

In future, it may be tried to propose a method to find more-for-less solutions of intuitionistic fuzzy/fuzzy linear/linear fractional transportation problems with mixed constraints based on these arithmetic operations.

References

1. A. Mishra, A. Kumar, Aggregation operators for various extensions of fuzzy set and its applications in transportation problems, in *Studies in Fuzziness and Soft Computing* (Springer Nature, Singapore, 2021)
2. V. Vidhya, P. Uma Maheswari, K. Ganesan, An alternate method for finding more for less solution to fuzzy transportation problem with mixed constraints. Soft Comput. **25**, 11989–11996 (2021)

MIX
Papier aus verantwortungsvollen Quellen
Paper from responsible sources
FSC® C105338

If you have any concerns about our products,
you can contact us on
ProductSafety@springernature.com

In case Publisher is established outside the EU,
the EU authorized representative is:
Springer Nature Customer Service Center GmbH
Europaplatz 3, 69115 Heidelberg, Germany

Printed by Libri Plureos GmbH
in Hamburg, Germany